T0338560

Bloch-Type Periodic Functions

Theory and Applications to Evolution Equations

SERIES ON CONCRETE AND APPLICABLE MATHEMATICS

ISSN: 1793-1142

Series Editor: Professor George A. Anastassiou
Department of Mathematical Sciences
University of Memphis
Memphis, TN 38152, USA

Published*

Vol. 13 Problems in Probability, Second Edition
by T. M. Mills

Vol. 14 Evolution Equations with a Complex Spatial Variable
by Ciprian G. Gal, Sorin G. Gal & Jerome A. Goldstein

Vol. 15 An Exponential Function Approach to Parabolic Equations
by Chin-Yuan Lin

Vol. 16 Frontiers in Approximation Theory
by George A. Anastassiou

Vol. 17 Frontiers in Time Scales and Inequalities
by George A. Anastassiou

Vol. 18 Differential Sheaves and Connections: A Natural Approach to Physical
Geometry
by Anastasios Mallios & Elias Zafiris

Vol. 19 Stochastic Models with Applications to Genetics, Cancers, AIDS and
Other Biomedical Systems (Second Edition)
by Wai-Yuan Tan

Vol. 20 Discrete Approximation Theory
by George A. Anastassiou & Merve Kester

Vol. 21 Functional Equations and Inequalities: Solutions and Stability Results
by John Michael Rassias, E. Thandapani, K. Ravi &
B. V. Senthil Kumar

Vol. 22 Bloch-Type Periodic Functions: Theory and Applications to
Evolution Equations
by Yong-Kui Chang, Gaston Mandata N'Guérékata & Rodrigo Ponce

*To view the complete list of the published volumes in the series, please visit:
http://www.worldscientific/series/scaam

Series on Concrete and Applicable Mathematics – Vol. 22

Bloch-Type Periodic Functions

Theory and Applications to Evolution Equations

Yong-Kui Chang
Xidian University, China

Gaston M. N'Guérékata
Morgan State University, USA

Rodrigo Ponce
Universidad de Talca, Chile

World Scientific

NEW JERSEY · LONDON · SINGAPORE · BEIJING · SHANGHAI · HONG KONG · TAIPEI · CHENNAI · TOKYO

Published by

World Scientific Publishing Co. Pte. Ltd.

5 Toh Tuck Link, Singapore 596224

USA office: 27 Warren Street, Suite 401-402, Hackensack, NJ 07601

UK office: 57 Shelton Street, Covent Garden, London WC2H 9HE

Library of Congress Cataloging-in-Publication Data

Names: Chang, Yong-Kui, author. | N'Guérékata, Gaston M., 1953–   author. |
    Ponce, Rodrigo (Professor of mathematics), author.
Title: Bloch-type periodic functions : theory and applications to evolution equations /
    Yong-Kui Chang, Xidian University, China; Gaston M. N'Guérékata,
    Morgan State University, USA; Rodrigo Ponce, Universidad de Talca, Chile.
Description: New Jersey : World Scientific, [2022] | Series: Series on concrete and applicable
    mathematics, 1793-1142 ; vol. 22 | Includes bibliographical references and index.
Identifiers: LCCN 2022006842 | ISBN 9789811254352 (hardcover) |
    ISBN 9789811254369 (ebook for institutions) | ISBN 9789811254376 (ebook for individuals)
Subjects: LCSH: Periodic functions. | Bloch constant.
Classification: LCC QA353.P4 .C43 2022 | DDC 515/.94--dc23/eng20220422
LC record available at https://lccn.loc.gov/2022006842

British Library Cataloguing-in-Publication Data
A catalogue record for this book is available from the British Library.

Cover image:

Glaciar Grey, Torres del Paine National Park, Chile, January 25, 2017

Photographed by Rodrigo Ponce

For photocopying of material in this volume, please pay a copying fee through the Copyright Clearance Center, Inc., 222 Rosewood Drive, Danvers, MA 01923, USA. In this case permission to photocopy is not required from the publisher.

For any available supplementary material, please visit
https://www.worldscientific.com/worldscibooks/10.1142/12780#t=suppl

Desk Editors: Soundararajan Raghuraman/Lai Fun Kwong

Typeset by Stallion Press
Email: enquiries@stallionpress.com

Printed in Singapore

Preface

The aim of this monograph is to give for the first time a unified and homogenous presentation of recent works on the theory of Bloch-type functions and their applications to evolution equations. The concept of Bloch functions goes back to the Swiss physicist Félix Bloch while working on the conductivity of crystalline solids (see for instance Bloch, 1929). Bloch functions generalize both periodic and anti periodic functions. They can be expressed in the form

$$\Psi(r) = e^{ik \cdot r} \psi(r),$$

where r denotes the position, Ψ the wave function, ψ a periodic function and k the wave vector. Since the paper by Hasler and N'Guérékata (2014) the study of periodic functions of Bloch-type has aroused great interest due to their importance in quantum physics and other branches of mathematical physics.

This book consists of nine chapters and an appendix. Chapter 1 is concerned with some preliminary facts. In Chapter 2, we introduce some new notions of generalized Bloch-type periodic functions and present some fundamental properties on spaces of such functions. In Chapter 3, we investigate the existence and uniqueness of generalized Bloch-type periodic solutions to semilinear integrodifferential equations with mixed kernel via a uniformly exponential stable resolvent operator family. Chapter 4 is devoted to establish some existence results for generalized Bloch-type periodic solutions to multi-term fractional evolution equations via a uniformly integrable resolvent operator family. Chapter 5 is mainly concerned with the existence and uniqueness of generalized Bloch-type periodic solutions to semilinear fractional evolution equations of degenerate type via the uniform integrability of a well-defined resolvent operator family. In Chapter 6, we show the existence and uniqueness of generalized Bloch-type periodic solutions to semilinear fractional integrodifferential equations via a uniformly integrable growth of a corresponding resolvent operator family. In Chapter 7, we establish some existence of pseudo S-asymptotically Bloch type periodic solutions to damped evolution equations with local or nonlocal initial conditions on the nonnegative real axis via suitable regularized families. Chapter 8 is focused upon some existence results for pseudo S-asymptotically Bloch-type periodic solutions to partial integrodifferential equations with local or nonlocal initial conditions via a uniformly exponentially stable resolvent operator family. In Chapter 9, we present some existence results of generalized Bloch-type solutions to semilinear integral equations via asymptotic decay of an integral resolvent family. The final is an appendix, which mainly includes norm continuity and characterization of compactness for fractional resolvent operator families appearing in Chapters 4 and 5, and applications to asymptotic behavior of solutions to abstract fractional Cauchy problems in the Caputo and Riemann–Liouville fractional derivatives, respectively.

The content of this monograph includes some new and unpublished results. It is useful for graduate students and researchers as seminar topics, graduate courses and reference book in *Pure and Applied Mathematics, Physics and Engineering*.

About the Authors

Dr Yong-Kui Chang is now working as a full professor in the School of Mathematics and Statistics, Xidian University, Xi'an, China. His main research interests include Bloch periodicity, almost periodicity and almost automorphy with applications to abstract evolution equations, functional differential equations and inclusion with applications, evolution systems and controls.

Dr Gaston Mandata N'Guérékata is a University Distinguished Professor and Associate Dean at Morgan State University in Baltimore, Maryland, USA. He received his college education from the University of Montreal in Canada. He is an American Mathematical Society (AMS) Fellow, a The World Academy of Sciences (TWAS) Fellow, an African Academy of Sciences (AAS) Fellow and author of over 280 publications including 11 books at the graduate/research level, some of them are cornerstones on the subjects. His contributions range from abstract harmonic analysis to almost periodicity, almost automorphy, fractional calculus and evolution equations. Dr N'Guérékata is also on the Editorial Boards of over 20 international journals.

Professor Rodrigo Ponce received his PhD in Mathematics at the University of Santiago de Chile, Chile, in 2011. He has published more than 40 scientific articles in the area of functional analysis, mainly on evolutionary equations, maximal regularity on UMD spaces, theory of semigroups of linear operators, theory of operators, resolvent families in Banach spaces, differential and integral equations, and fractional differential equations. Currently, he works as an Assistant Professor at the Institute of Mathematics of the University of Talca, Chile, where he teaches mathematics in undergraduate and graduate courses, and collaborates in the Mathematics Olympiad and in the training of secondary school teachers in his country.

Acknowledgments

Yong-Kui Chang would like to thank his graduate students Zhi-Han Zhao, Rui Zhang, Jian-Qiong Zhao, Xiao-Xia Luo, Zhuan-Xia Cheng, Yan-Tao Bian, Chao Tang, Xue-Yan Wei, Mei-Juan Zhang, Shan Zheng, Yanyan Wei, Siqi Chen, for their help and collaboration. He also acknowledges with gratitude the support of NSF of China, the Key Project of Chinese Ministry of Education, China Postdoctoral Science Foundation funded project, Program for New Century Excellent Talents in University, Program for Longyuan Youth Innovative Talents of Gansu Province of China, and NSF of Gansu Province of China, NSF of Shaanxi Province of China during his working.

G. M. N'Guérékata would like to thank professor Maximillian Hasler and his students Darin Brindle and Roger Enock Oueama-Guengai for their collaboration.

R. Ponce would like to thank his mentor professor C. Lizama, and to his graduate students A. Pereira and S. Rueda for their help and collaboration. He also thanks the support of Conicyt-ANID of Chile (11130619).

We are thankful to professors B. de Andrade, E. Cuesta, C. Cuevas, H. R. Henríquez, E. Hernández, C. Lizama, M. Pierri for their exceptional contribution to theories of resolvent operator families, damped evolution equations and S-asymptotic periodicity. We also tender our thanks to anonymous referees for carefully reading the manuscript and giving valuable comments to improve this book. Meanwhile, we express our thanks to series editor professor George A. Anastassiou, and the editorial members of World Scientific Publishing Co. Pte. Ltd. Kwong Lai Fun and Soundararajan T. R. for their cooperation during the preparation of this book for publication. Finally, we are deeply indebted to our beloved family members.

Yong-Kui Chang,
Gaston Mandata N'Guérékata,
Rodrigo Ponce

Contents

Preface   v

About the Authors   vii

Acknowledgments   ix

1. Preliminaries   1

2. Bloch-Type Periodic Functions and Generalizations   13

    2.1   Bloch-Type Periodic Functions   14
    2.2   Pseudo Bloch-Type Periodic Functions   17
    2.3   Weighted Pseudo Bloch-Type Periodic Functions   20
    2.4   *S*-Class Asymptotically Bloch-Type Periodic
          Functions   31
    2.5   Stepanov-Like Asymptotically Bloch-Type Periodic
          Functions   47

3. Bloch-Type Periodic Solutions to Semilinear
   Integrodifferential Equations of Mixed Kernel   61

    3.1   Uniform Exponential Stability of Solutions to a Volterra
          Equation   62
    3.2   Linear Integrodifferential Equation   67
    3.3   Semilinear Integrodifferential Equation   72

4. Bloch-Type Periodic Solutions to Multi-Term Fractional
   Evolution Equations   79

    4.1   Asymptotic Behavior and Uniform Integrability of the
          Resolvent Family   81

4.2   Linear Fractional Evolution Equation   87
4.3   Semilinear Fractional Evolution Equation   90

5. Bloch-Type Periodic Solutions to Fractional Evolution
    Equations of Sobolev Type   99

    5.1   Asymptotic Behavior of Sobolev-Type Resolvent
          Family 100
    5.2   Bloch-Type Periodic Solutions 105

6. Bloch-Type Periodic Solutions to Fractional
    Integrodifferential Equations   113

    6.1   Fractional Integrodifferential Equation in the Linear
          Case 114
    6.2   Fractional Integrodifferential Equation in the Semilinear
          Case 121

7. Asymptotically Bloch-Type Periodic Solutions to
    Damped Evolution Equations   129

    7.1   Some Basic Results 130
    7.2   Asymptotically Bloch-Type Periodic Solutions 134

8. Asymptotically Bloch-Type Periodic Solutions to Partial
    Integrodifferential Equations   145

    8.1   Some Basic Results 146
    8.2   Asymptotically Bloch-Type Periodic Solutions 147

9. Bloch-Type Periodic Solutions to Semilinear Integral Equations 153

    9.1   Integral Resolvent Family and Linear Integral
          Equation 153
    9.2   Semilinear Integral Equation 156

Appendix A   Compactness of Fractional Resolvent Operator
    Families   165

    A.1   Norm Continuity and Compactness of the Resolvent
          Operator $S_{\alpha,\beta}(t)$ 165

A.2   Norm Continuity and Compactness of the Resolvent
        Operator $S_{\alpha,\beta}^{E}(t)$ 172
A.3   Applications 178

Bibliography   183

Index   193

Chapter 1

Preliminaries

Let $\mathbb{R}$ ($\mathbb{C}$) be the set of all real (complex) numbers and $\mathbb{R}_+$ be the collection of all nonnegative real numbers. Let $\mathbb{N}$ denote the set of all positive integers and $\mathbb{C}_+ = \{\lambda \in \mathbb{C} : \operatorname{Re}\lambda > 0\}$. Let $(X, \|\cdot\|)$ be a Banach space and $C(\mathbb{R}, X)$ ($C(\mathbb{R}_+, X)$) denotes the space of all continuous functions $f : \mathbb{R} \to X$ ($f : \mathbb{R}_+ \to X$). The notation $C_0(\mathbb{R}_+, X)$ is the space of all continuous functions $g : \mathbb{R}_+ \to X$ such that $\lim_{t\to\infty} \|g(t)\| = 0$. Let $BC(\mathbb{R}, X)$ ($BC(\mathbb{R}_+, X)$) stand for the Banach space formed by all bounded continuous functions $f : \mathbb{R} \to X$ ($f : \mathbb{R}_+ \to X$) with sup-norm $\|f\|_\infty = \sup_{t\in\mathbb{R}} \|f(t)\|$ ($\|f\|_\infty = \sup_{t\in\mathbb{R}_+} \|f(t)\|$). The space $BC(\mathbb{R} \times X, X)$ ($BC(\mathbb{R}_+ \times X, X)$) is the set consisting of all functions $f : \mathbb{R} \times X \to X$ ($f : \mathbb{R}_+ \times X \to X$) such that $f(\cdot, x) \in BC(\mathbb{R}, X)$ ($f(\cdot, x) \in BC(\mathbb{R}_+, X)$) uniformly for each x in any compact subset of X. We denote by $\mathcal{B}(X, Y)$ the space of all bounded linear operators from X to Y and will be abbreviated by $\mathcal{B}(X)$ if $X = Y$.

Let $\mathbb{I}$ be any subset of $\mathbb{R}$. A function $f : \mathbb{I} \to X$ is measurable if there exists a sequence of simple functions g_n such that

$$f(t) = \lim_{n\to\infty} g_n(t),$$

for almost all $t \in \mathbb{I}$, and a function $f : \mathbb{I} \to X$ is simple if it has the form

$$f(t) = \sum_{k=1}^{n} x_k \chi_{\Omega_k}(t),$$

for some $n \in \mathbb{N}, x_k \in X$ and Lebesgue measurable sets $\Omega_k \subset \mathbb{I}$ with finite Lebesgue measure $m(\Omega_k)$, where χ_Ω denotes the characteristic function of Ω. The notation $L^p(\mathbb{I}, X)$ ($1 \leq p < \infty$) denotes the space of X-valued

Bochner integrable functions defined on $\mathbb{I}$ with the norm

$$\|x\|_p := \|x\|_{L^p(\mathbb{I},X)} = \left(\int_{\mathbb{I}} \|x(t)\|^p dt \right)^{1/p}.$$

We denote $L^p_{\mathrm{loc}}(\mathbb{R}, X)$ by the space of all pth locally Bochner integrable functions. For a simple function $g : \mathbb{I} \to X$, $g = \Sigma_{i=1}^m x_i \chi_{\Omega_i}$, we define

$$\int_{\mathbb{I}} g(t)dt := \sum_{i=1}^m x_i m(\Omega_i).$$

It is easy to see that the above defined integral is independent of the expression $g = \Sigma_{i=1}^m x_i \chi_{\Omega_i}$, and it is linear. A function $f : \mathbb{I} \to X$ is said to be Bochner integrable if there exist simple functions g_n such that $g_n \to f$ pointwise a.e., and

$$\lim_{n \to \infty} \int_{\mathbb{I}} \|f(t) - g_n(t)\| dt = 0.$$

If $f : \mathbb{I} \to X$ is Bochner integrable, then the Bochner integral of f on $\mathbb{I}$ is defined by

$$\int_{\mathbb{I}} f(t)dt := \lim_{n \to \infty} \int_{\mathbb{I}} g_n(t)dt.$$

It can be checked that this limit exists and is independent of the choice of the sequence $\{g_n\}$. When $\mathbb{I}$ is a rectangle, we denote a Bochner integral by $\int_{\mathbb{I}} f(s,t)d(s,t)$.

Lemma 1.1 (Pettis theorem, Arendt *et al.*, 2001; N'Guérékata, 2005). *A function $f : \mathbb{I} \to X$ is measurable if and only if the following conditions hold:*

(a) *f is weakly measurable, i.e., for each $x^* \in X^*$, the dual space of X, the function $(x^* f)(t) : \mathbb{I} \to \mathbb{R}$ is measurable.*
(b) *f is almost separably valued, i.e., there exists a set $\mathbb{S} \subset \mathbb{I}$ of measure 0 such that $f(\mathbb{I} \setminus \mathbb{S})$ is separable.*

Lemma 1.2 (Bochner theorem, Arendt *et al.*, 2001; N'Guérékata, 2005). *A measurable function $f : \mathbb{I} \to X$ is Bochner integrable if and only if $\|f\|$ is Lebesgue integrable. Furthermore, if f is Bochner integrable, then*

$$\left\| \int_{\mathbb{I}} f(t)dt \right\| \leq \int_{\mathbb{I}} \|f(t)\| dt.$$

Lemma 1.3 (Arendt *et al.*, 2001). *Let* $\mathcal{T} : X \to Y$ *be a bounded linear operator between Banach spaces X and Y, and let $f : \mathbb{I} \to X$ be Bochner integrable. Then $\mathcal{T} \circ f : t \mapsto \mathcal{T}(f(t))$ is Bochner integrable and*

$$\mathcal{T} \int_{\mathbb{I}} f(t)dt = \int_{\mathbb{I}} \mathcal{T}(f(t))dt.$$

Lemma 1.4 (Arendt *et al.*, 2001). *Let A be a closed linear operator on X. Let $f : \mathbb{I} \to X$ be Bochner integrable. Suppose that $f(t) \in D(A)$ (the domain of A) for all $t \in \mathbb{I}$ and $A \circ f : \mathbb{I} \to X$ is Bochner integrable. Then*

$$\int_{\mathbb{I}} f(t)dt \in D(A), \quad and \quad A \int_{\mathbb{I}} f(t)dt = \int_{\mathbb{I}} A(f(t))dt.$$

Lemma 1.5 (Dominated convergence theorem, Arendt *et al.*, 2001; N'Guérékata, 2005). *Let $f_n : \mathbb{I} \to X$ ($n \in \mathbb{N}$) be Bochner integrable functions. Assume that:*

(i) *There exists an integrable function $g : \mathbb{I} \to \mathbb{R}$ such that*

$$\|f_n(t)\| \le g(t)$$

a.e. on $\mathbb{I}$ for all $n \in \mathbb{N}$.
(ii) *$f(t) := \lim_{n \to \infty} f_n(t)$ exists a.e. on $\mathbb{I}$.*

Then f is Bochner integrable and

$$\int_{\mathbb{I}} f(t)dt = \lim_{n \to \infty} \int_{\mathbb{I}} f_n(t)dt.$$

Furthermore,

$$\int_{\mathbb{I}} \|f(t) - f_n(t)\|dt \to 0 \ as \ n \to \infty.$$

Lemma 1.6 (Fubini theorem, Arendt *et al.*, 2001). *Let $\mathbb{I} = \mathbb{I}_1 \times \mathbb{I}_2$ be a rectangle in $\mathbb{R}^2$. Let $f : \mathbb{I} \to X$ be measurable, and suppose that*

$$\int_{\mathbb{I}_1} \int_{\mathbb{I}_2} \|f(s, t)\|dtds < \infty.$$

Then f is Bochner integrable and the repeated integrals

$$\int_{\mathbb{I}_1} \int_{\mathbb{I}_2} f(s, t)dtds, \quad \int_{\mathbb{I}_2} \int_{\mathbb{I}_1} f(s, t)dsdt$$

exist and are equal, and they coincide with the double integral

$$\int_{\mathbb{I}} f(s, t)d(s, t).$$

Lemma 1.7 (Mean value theorem for the Bochner integral, Diestel and Uhl, Jr., 1977). *Let f be Bochner integrable with respect to measure μ. Then for each $\Omega \in \Sigma$ field with $\mu(\Omega) > 0$, one has*

$$\frac{1}{\mu(\Omega)} \int_{\Omega} f d\mu \in \overline{co(f(\Omega))},$$

where $\overline{co(\mathbb{E})}$ denotes the closed convex hull of $\mathbb{E}$.

For a linear operator A defined on X, $D(A), \sigma(A), \rho(A)$ denote the domain, spectrum and resolvent set of A, respectively. Particularly, $\sigma_p(A)$ denotes the point spectrum of A. The family $R(\lambda, A) = (\lambda I - A)^{-1}, \lambda \in \rho(A)$ of bounded linear operators is called the resolvent of A, where I is the identity operator on X. We list the following definitions and results.

Definition 1.1 (N'Guérékata, 2005, 2021). Let $A : X \to X$ be a linear operator with $D(A) \subseteq X$, in a Banach space X. The family $\{T(t)\}_{t \in \mathbb{R}_+}$ of bounded linear operators on X is said to be a C_0-semigroup (or strongly continuous semigroup) if

1. For all $x \in X$, the mapping $T(t)x : \mathbb{R}_+ \to X$ is continuous.
2. For all $s, t \in \mathbb{R}_+$, $T(s + t) = T(s)T(t)$.
3. $T(0) = I$.

The operator A is called the infinitesimal generator (or generator in short) of the C_0-semigroup $T(t)$ if

$$Ax = \lim_{t \to 0^+} \frac{T(t)x - x}{t},$$

where

$$D(A) = \left\{ x \in X | \lim_{t \to 0^+} \frac{T(t)x - x}{t} \text{ exists} \right\}.$$

It is observed that A commutes with $T(t)$ on $D(A)$. We also have the following properties.

Lemma 1.8 (N'Guérékata, 2005, 2021). *If $\{T(t)\}_{t \in \mathbb{R}_+}$ is a C_0-semigroup, then:*

(1) *The function $t \mapsto \|T(t)\|, \mathbb{R}_+ \to \mathbb{R}_+$ is measurable and bounded on any compact interval of $\mathbb{R}_+$.*
(2) *The domain $D(A)$ of its generator is dense in X.*
(3) *The generator A is a closed linear operator.*

Definition 1.2 (Mátrai, 2008). A C_0-semigroup $\{T(t)\}_{t\geq 0}$ is eventually norm continuous if there exists a $t_0 \geq 0$ such that the map $t \mapsto T(t)$ is continuous with respect to the uniform operator topology for $t > t_0$. The semigroup is immediately norm continuous if $t_0 = 0$ can be chosen.

In a Hilbert space, we have the following characterization of immediate and eventual norm continuity for a C_0-semigroup.

Lemma 1.9 (Mátrai, 2008). *Let A be the generator of a C_0-semigroup $\{T(t)\}_{t\geq 0}$ on a Hilbert space $(H, \|\cdot\|)$. Let $s(A) = \sup\{Re\lambda : \lambda \in \sigma(A)\}$ denote the spectral bound of A. Then*

(a) *$\{T(t)\}_{t\geq 0}$ is immediately norm continuous, i.e., the mapping $T : (0, \infty) \to \mathcal{B}(H)$ is continuous if and only if for some $\mu_0 > s(A)$,*

$$\lim_{\mu\in\mathbb{R}, |\mu|\to\infty} \|R(\mu_0 + i\mu, A)\| = 0. \tag{1.1}$$

(b) *$\{T(t)\}_{t\geq 0}$ is eventual norm continuity, i.e., there exists a $t_0 > 0$ such that the mapping $T : (t_0, \infty) \to \mathcal{B}(H)$ is continuous, if and only if there exists an $n \in \mathbb{N}$ and a $\mu_0 > s(A)$ such that*

$$\lim_{\mu\in\mathbb{R}, |\mu|\to\infty} \|R(\mu_0 + i\mu, A)^n T(t_0)\| = 0. \tag{1.2}$$

Moreover, conditions (1.1) and (1.2) are necessary for immediate and eventual norm continuity in arbitrary Banach spaces.

Definition 1.3 (Lizama and N'Guérékata, 2010). A strongly continuous family $\{T(t)\}_{t\geq 0} \subseteq \mathcal{B}(X)$ is said to be uniformly integrable (or 1-integrable) if it verifies

$$\int_0^\infty \|T(t)\| dt < \infty.$$

Definition 1.4 (N'Guérékata, 2005, 2021). A strongly continuous family $\{T(t)\}_{t\geq 0} \subseteq \mathcal{B}(X)$ is said to be of type (M, δ) or exponentially bounded if there exist constants $M > 0$ and $\delta \in \mathbb{R}$ such that $\|T(t)\| \leq Me^{\delta t}$ for all $t \geq 0$.

Definition 1.5 (N'Guérékata, 2005, 2021). A strongly continuous family $\{T(t)\}_{t\geq 0} \subseteq \mathcal{B}(X)$ is said to be uniformly exponentially stable if there exist constants $M, \delta > 0$ such that $\|T(t)\| \leq Me^{-\delta t}$ for all $t \geq 0$.

Lemma 1.10 (Prüss, 1993). *Suppose $g : \mathbb{C}_+ \to X$ is holomorphic and satisfies*

$$|\lambda g(\lambda)| + |\lambda^2 g'(\lambda)| \leq M, \text{ for all } Re\lambda > 0,$$

where $M > 0$ is a constant. Then

$$m_\infty(g) := \sup\left\{\frac{\mu^{n+1}\|g^{(n)}(\mu)\|}{n!} : \mu > 0, n \in \mathbb{N} \cup \{0\}\right\} < \infty.$$

Lemma 1.11 (Engel and Nagel, 1999). *Let $(A, D(A))$ be a linear operator on a Banach space X and let $\delta \in \mathbb{R}, M \geq 1$ be constants. Then the following properties are equivalent:*

(a) *$(A, D(A))$ generates a strongly continuous semigroup $\{T(t)\}_{t \geq 0}$ satisfying $\|T(t)\| \leq Me^{\delta t}$ for all $t \geq 0$ and for some constants $M > 0$ and $\delta \in \mathbb{R}$.*

(b) *$(A, D(A))$ is closed, densely defined, and for every $\lambda > \delta$ one has $\lambda \in \rho(A)$ and*

$$\|[(\lambda - \delta)R(\lambda, A)]^n\| \leq M, \quad \text{for all } n \in \mathbb{N}.$$

(c) *$(A, D(A))$ is closed, densely defined, and for every complex λ with $\mathrm{Re}\lambda > \delta$ one has $\lambda \in \rho(A)$ and*

$$\|R(\lambda, A)^n\| \leq \frac{M}{(\mathrm{Re}\lambda - \delta)^n}, \quad \text{for all } n \in \mathbb{N}.$$

Lemma 1.12 (Arendt et al., 2001). *Let $f \in L^1_{\mathrm{loc}}(\mathbb{R}_+, X)$ and let $T : \mathbb{R}_+ \to \mathcal{B}(X, Y)$ be strongly continuous. Then the convolution*

$$(T * f)(t) := \int_0^t T(t - s)f(s)ds$$

exists (as a Bochner integral) and defines a continuous function $T * f : \mathbb{R}_+ \to Y$.

Definition 1.6 (Arendt et al., 2001). A closed linear operator $(A, D(A))$ with dense domain $D(A)$ in a Banach space X is said to be sectorial (of angle θ) if there exists $0 < \theta \leq \pi/2$ such that the sector

$$\sum_\theta := \left\{\lambda \in \mathbb{C} : |\arg(\lambda)| < \frac{\pi}{2} + \theta\right\} \setminus \{0\}$$

is contained in the resolvent set $\rho(A)$, and if for each $\varepsilon \in (0, \theta)$ there exists $M_\varepsilon \geq 1$ such that

$$\|R(\lambda, A)\| \leq \frac{M_\varepsilon}{|\lambda|}, \quad \text{for all } 0 \neq \lambda \in \overline{\sum_{\theta-\varepsilon}}.$$

For more details on semigroups of linear operators with applications, one can refer to Engel and Nagel (1999) and Pazy (1983).

We recall that the Laplace transform of a function $f \in L^1_{\text{loc}}(\mathbb{R}_+, X)$ is given by

$$\mathcal{L}(f)(\lambda) := \hat{f}(\lambda) := \int_0^\infty e^{-\lambda t} f(t) dt, \quad \text{Re}\lambda > \varpi,$$

where the integral is absolutely convergent for $\text{Re}\lambda > \varpi$. For $f : \mathbb{R}_+ \to X$, the abscissa of convergence of $\hat{f}$ is given by

$$\text{abs}(f) := \inf\{\text{Re}\lambda : \hat{f}(\lambda) \text{ exists}\},$$

and the exponential growth bound is given by

$$\varpi(f) := \left\{\varpi \in \mathbb{R} : \sup_{t \geq 0} \|e^{-\varpi t} f(t)\| < \infty\right\}.$$

A function f is said to be Laplace transformable if $\text{abs}(f) < \infty$. It has the fact that

$$\text{abs}(f) \leq \text{abs}(\|f\|) \leq \varpi(f).$$

We list the following basic results.

Lemma 1.13 (Arendt *et al.*, 2001). *Let* $f \in L^1_{\text{loc}}(\mathbb{R}_+, X)$. *Then we have:*

(1) *If* $\text{abs}(f) < \infty$, *then* $\lambda \mapsto \hat{f}(\lambda)$ *is holomorphic for* $\text{Re}\lambda > \text{abs}(f)$ *and, for all* $n \in \mathbb{N} \cup \{0\}$ *and* $\text{Re}\lambda > \text{abs}(f)$,

$$\hat{f}^{(n)}(\lambda) = \int_0^\infty e^{-\lambda t}(-t)^n f(t) dt.$$

(2) *Let* $\lambda \in \mathbb{C}$. *Then for* $g(t) := e^{-\mu t} f(t), t \in \mathbb{R}_+, \mu \in \mathbb{C}$, $\hat{g}(\lambda)$ *exists if and only if* $\hat{f}(\lambda + \mu)$ *exists, and*

$$\hat{g}(\lambda) = \hat{f}(\lambda + \mu);$$

For $f_s(t) := f(s + t), t, s \in \mathbb{R}_+$, $\widehat{f_s}(\lambda)$ *exists if and only if* $\hat{f}(\lambda)$ *exists, and*

$$\widehat{f_s}(\lambda) = e^{\lambda s}\left(\hat{f}(\lambda) - \int_0^s e^{-\lambda t} f(t) dt\right).$$

Meanwhile, for

$$h_s(t) := \begin{cases} f(t - s), & t - s \geq 0, \\ 0, & 0 \leq t < s, \end{cases}$$

$\widehat{h_s}(\lambda)$ *exists if and only if* $\hat{f}(\lambda)$ *exists, and*

$$\widehat{h_s}(\lambda) = e^{-\lambda s} \hat{f}(\lambda).$$

(3) *Let $\mathcal{T} \in \mathcal{B}(X, Y)$ and $(\mathcal{T} \circ f)(t) = \mathcal{T}(f(t))$. Then $\mathcal{T} \circ f \in L^1_{\text{loc}}(\mathbb{R}_+, Y)$. If $\hat{f}(\lambda)$ exists, then $\widehat{\mathcal{T} \circ f}(\lambda)$ exists and*

$$\widehat{\mathcal{T} \circ f}(\lambda) = \mathcal{T}(\hat{f}(\lambda)).$$

(4) *Let A be a closed linear operator on X. Suppose that $f(t) \in D(A)$ a.e. and $A \circ f \in L^1_{\text{loc}}(\mathbb{R}_+, X)$. Let $\lambda \in \mathbb{C}$. If $\hat{f}(\lambda)$ and $\widehat{A \circ f}(\lambda)$ both exists, then $\hat{f}(\lambda) \in D(A)$ and*

$$\widehat{A \circ f}(\lambda) = A(\hat{f}(\lambda)).$$

(5) *Let $k \in L^1_{\text{loc}}(\mathbb{R}_+), \lambda \in \mathbb{C}$. If $Re\lambda > \max\{abs(|k|), abs(f)\}$, then $\widehat{k * f}(\lambda)$ exists and*

$$\widehat{k * f}(\lambda) = \hat{k}(\lambda)\hat{f}(\lambda).$$

(6) *Let $F(t) = \int_0^t f(s)ds$. If $Re\lambda > 0$ and $\hat{f}(\lambda)$ exists, then $\widehat{F}(\lambda)$ exists and*

$$\widehat{F}(\lambda) = \hat{f}(\lambda)/\lambda.$$

(7) *Let $f : \mathbb{R}_+ \to X$ be absolutely continuous and differentiable a.e. If $Re\lambda > 0$ and $\widehat{f'}(\lambda)$ exists, then $\hat{f}(\lambda)$ exists and*

$$\widehat{f'}(\lambda) = \lambda \hat{f}(\lambda) - f(0).$$

For detailed results on vector valued Laplace transforms, please refer to Arendt *et al.* (2001).

We now recall some basic results on abstract Volterra equations which are taken from Lizama (2000) and Prüss (1993). Consider the following Volterra equation of scalar type defined on a complex Banach space X

$$u(t) = f(t) + \int_0^t a(t - s)Au(s)ds, \quad t \in [0, b], \tag{1.3}$$

where $b > 0$, A is a closed linear unbounded operator, and $a(\cdot) \in L^1_{\text{loc}}(\mathbb{R}_+)$ is a scalar kernel $\neq 0$, $f \in C([0, b], X)$.

Definition 1.7 (Prüss, 1993). A family $\{\mathcal{S}(t)\}_{t \geq 0} \subseteq \mathcal{B}(X)$ is called a resolvent for Eq. (1.3) if the following conditions are satisfied:

(S-1) $\mathcal{S}(t)$ is strongly continuous on $\mathbb{R}_+$ and $\mathcal{S}(0) = I$.
(S-2) $\mathcal{S}(t)$ commutes with A, which means that $\mathcal{S}(t)D(A) \subseteq D(A)$ and $A\mathcal{S}(t)x = \mathcal{S}(t)Ax$ for all $x \in D(A)$ and $t \geq 0$.
(S-3) The following resolvent equation holds

$$\mathcal{S}(t)x = x + \int_0^t a(t - s)A\mathcal{S}(s)x ds$$

for all $x \in D(A), t \geq 0$.

Lemma 1.14 (Prüss, 1993). *Equation* (1.3) *is well posed if and only Eq.* (1.3) *admits a resolvent* $S(t)$. *If this is the case we have* $a*S(t)x \in D(A)$ *for all* $x \in X, t \geq 0$ *and*

$$S(t)x = x + A \int_0^t a(t-s)S(s)xds, \ x \in X, t \geq 0.$$

Particularly, $Aa * S$ *is strongly continuous in* X.

Lemma 1.15 (Prüss, 1993). *Let* A *be a closed linear unbounded operator in the space* X *with dense domain* $D(A)$ *and let* $a \in L^1_{loc}(\mathbb{R}_+)$ *with*

$$\int_0^\infty e^{-\delta t}|a(t)|dt < \infty.$$

Then (1.3) *admits a resolvent* $S(t)$ *of type* (M, δ) *if and only if the following conditions hold:*

(a) $\hat{a}(\lambda) \neq 0$ *and* $1/\hat{a}(\lambda) \in \rho(A)$ *for all* $\lambda > \delta$.
(b) $H(\lambda) := (I - \hat{a}(\lambda)A)^{-1}/\lambda$ *satisfies the estimates*

$$\|H^{(n)}(\lambda)\| \leq \frac{Mn!}{(\lambda - \delta)^{n+1}}, \quad \text{for all } \lambda > \delta \text{ and } n \in \mathbb{N} \cup \{0\}.$$

Definition 1.8 (Lizama, 2000). *Let* $k(\cdot) \in C(\mathbb{R}_+)$ *be a scalar kernel. A family* $\{\mathcal{R}(t)\}_{t \geq 0} \subseteq \mathcal{B}(X)$ *is called a* k-*regularized resolvent for Eq.* (1.3) *if the following conditions are satisfied:*

(R-1) $\mathcal{R}(t)$ *is strongly continuous on* $\mathbb{R}_+$ *and* $\mathcal{R}(0) = k(0)I$.
(R-2) $\mathcal{R}(t)x \in D(A)$ *and* $A\mathcal{R}(t)x = \mathcal{R}(t)Ax$ *for all* $x \in D(A)$ *and* $t \geq 0$.
(R-3) *The following* k-*regularized resolvent equation holds*

$$\mathcal{R}(t)x = k(t)x + \int_0^t a(t-s)A\mathcal{R}(s)xds$$

for all $x \in D(A), t \geq 0$.

Lemma 1.16 (Lizama, 2000). *Assume that Eq.* (1.3) *admits a* k-*regularized resolvent. Then* $a * \mathcal{R}(t)x \in D(A)$ *for all* $x \in X, t \geq 0$ *and*

$$\mathcal{R}(t)x = k(t)x + A \int_0^t a(t-s)\mathcal{R}(s)xds, \quad x \in X, t \geq 0.$$

Lemma 1.17 (Lizama, 2000). *Let* $\mathcal{R}(t)$ *be an exponentially bounded and strongly continuous operator family in* $\mathcal{B}(X)$ *such that the Laplace transform* $\widehat{\mathcal{R}}(\lambda)$ *exists for* $\lambda > \varpi$. *Then* $\{\mathcal{R}(t)\}_{t \geq 0}$ *is a* k-*regularized resolvent for Eq.* (1.3) *if and only if for every* $\lambda > \varpi$, $(I - \hat{a}(\lambda)A)^{-1}$ *exists in* $\mathcal{B}(X)$ *and*

$$\hat{k}(\lambda)(I - \hat{a}(\lambda)A)^{-1}x = \int_0^\infty e^{-\lambda s}\mathcal{R}(s)xds, \quad \text{for all } x \in X.$$

Lemma 1.18 (Lizama, 2000). *Let A be a closed linear densely defined operator on a Banach space X. Then Eq. (1.3) admits a k-regularized resolvent $\mathcal{R}(t)$ of type (M, δ) if and only if the following conditions hold:*

(a) *$\hat{a}(\lambda) \neq 0$ and $1/\hat{a}(\lambda) \in \rho(A)$ for all $\lambda > \delta$.*
(b) *$H(\lambda) := \hat{k}(\lambda)(I - \hat{a}(\lambda)A)^{-1}$ satisfies the estimates*

$$\|H^{(n)}(\lambda)\| \leq \frac{Mn!}{(\lambda - \delta)^{n+1}}, \quad \text{for all } \lambda > \delta \text{ and } n \in \mathbb{N} \cup \{0\}.$$

We finally recall some compactness criteria and fixed point theorems which will be use in the sequel.

Let $h : \mathbb{R}_+ \to [1, \infty)$ be a continuous function which is nondecreasing and $h(t) \to \infty$ as $t \to \infty$. Let the space $C_h(X)$ be defined by

$$C_h(X) := \left\{ u \in C(\mathbb{R}_+, X) : \lim_{t \to \infty} \frac{\|u(t)\|}{h(t)} = 0 \right\}, \tag{1.4}$$

with the norm $\|u\|_h = \sup_{t \in \mathbb{R}_+} \frac{\|u(t)\|}{h(t)}$.

Lemma 1.19 (Cuevas and Henríquez, 2011). *A subset $\mathbb{K} \subseteq C_h(X)$ is relatively compact if it satisfies the following conditions:*

(c1) *The $\mathbb{K}_b = \left\{ u|_{[0,b]} : u \in \mathbb{K} \right\}$ is relatively compact in $C([0, b], X)$ for all $b \geq 0$.*
(c2) *$\lim_{t \to \infty} \frac{\|u(t)\|}{h(t)} = 0$ uniformly for all $u \in \mathbb{K}$.*

Let $\hbar : \mathbb{R} \to \mathbb{R}$ be a continuous function such that $\hbar(t) \geq 1$ for all $t \in \mathbb{R}$ and $\hbar(t) \to \infty$ as $|t| \to \infty$. We consider the space

$$C_\hbar(X) = \left\{ u \in C(\mathbb{R}, X) : \lim_{|t| \to \infty} \frac{\|u(t\|)}{\hbar(t)} = 0 \right\}, \tag{1.5}$$

with the norm $\|u\|_\hbar = \sup_{t \in \mathbb{R}} \frac{\|u(t)\|}{\hbar(t)}$.

Lemma 1.20 (Henríquez and Lizama, 2009). *A subset $\mathbb{K} \subseteq C_\hbar(X)$ is a relatively compact set if it verifies the following conditions:*

(c-1) *The set $\mathbb{K}(t) = \{u(t) : u \in \mathbb{K}\}$ is relatively compact in X for each $t \in \mathbb{R}$.*
(c-2) *The set $\mathbb{K}$ is equicontinuous.*
(c-3) *For each $\varepsilon > 0$ there exists $L > 0$ such that $\|u(t)\| \leq \varepsilon \hbar(t)$ for all $u \in \mathbb{K}$ and all $|t| > L$.*

Lemma 1.21 (Banach fixed point theorem, Granas and Dugundji, 2003). *Let (Z, d) be a complete metric space. If $\Upsilon : Z \to Z$ is a contraction, i.e., there exists a constant $L \in [0, 1)$ such that*

$$d(\Upsilon(x), \Upsilon(y)) \le Ld(x, y), \quad \text{for all } x, y \in Z,$$

then Υ has a unique fixed point in Z.

Corollary 1.1. *Let (Z, d) be a complete metric space and $\Upsilon : Z \to Z$. If there exist constants $L \in [0, 1)$ and $n_0 \in \mathbb{N}$ such that*

$$d(\Upsilon^{n_0}(x), \Upsilon^{n_0}(y)) \le Ld(x, y), \quad \text{for all } x, y \in Z,$$

then Υ has a unique fixed point in Z, where $\Upsilon^2 x = \Upsilon(\Upsilon x), \Upsilon^3 x = \Upsilon(\Upsilon^2 x), \dots, \Upsilon^{n_0} x = \Upsilon(\Upsilon^{n_0-1} x), \dots$.

Lemma 1.22 (Leray-Schauder alternative theorem, Granas and Dugundji, 2003). *Let $\mathbb{B}$ be a convex subset of a Banach space X. Suppose that $0 \in \mathbb{B}$. If $F : \mathbb{B} \to \mathbb{B}$ is a completely continuous map, then either F has a fixed point, or the set $\{x \in \mathbb{B} : x = \lambda F(x), 0 < \lambda < 1\}$ is unbounded.*

Lemma 1.23 (Krasnoselskii fixed point theorem, Granas and Dugundji, 2003). *Let $\mathbb{B}$ be a closed convex and nonempty subset of a Banach space X. Let Q_1 and Q_2 be two operators such that*

(i) *If $u, v \in \mathbb{B}$, then $Q_1 u + Q_2 v \in \mathbb{B}$.*
(ii) *Q_1 is a contraction.*
(iii) *Q_2 is compact and continuous.*

Then, there exists $z \in \mathbb{B}$ such that $z = Q_1 z + Q_2 z$.

Chapter 2

Bloch-Type Periodic Functions and Generalizations

It is well known that solutions to some equations from propagation of heat or waves in solid state physics, such as wave functions satisfying the Schödinger equation with the potential field having crystal lattice periodicity, often behave Bloch type periodicity (or (ω, k)-Bloch periodicity), see for instance (Bloch, 1929; Kittel, 2005; Ren, 2006). It can be found that Bloch type periodicity has classical ω-periodicity and ω-antiperiodicity as special cases. In Hasler and N'Guérékata (2014), Hasler and N'Guérékata established theories on the space of Bloch type periodic functions such as completeness, convolution and composite properties from the viewpoint of mathematics, and studied the existence and uniqueness of Bloch type periodic solutions to a semilinear integrodifferential equation in Banach spaces. The notion of asymptotically Bloch type periodic function was also presented in Hasler and N'Guérékata (2014), which can be seen as the Bloch type periodic function with a perturbation vanishing at infinity and a natural generalization of asymptotically ω-periodic function (see Gao *et al.*, 2006) and asymptotically ω-antiperiodic function (see Chen and Wang, 2015). Kostić and Velinov (2017) analyzed the existence and uniqueness of asymptotically Bloch type periodic solutions to an abstract fractional nonlinear differential inclusions with piecewise constant argument in Banach spaces. As is known, the ω-periodicity or ω-antiperiodicity have some other significant generalizations such as (weighted) pseudo ω-periodicity and (weighted) pseudo ω-antiperiodicity, which has been considerably applied to investigate the existence (and uniqueness) of (weighted) pseudo ω-periodic/ω-antiperiodic solutions to various evolution equations in Banach spaces, see for instance Al-Islam *et al.* (2012), Alvarez *et al.* (2015), Cao *et al.* (2012), Liu (2010) and Xia (2014) and references therein. On the other side, Henríquez, Pierri and Táboas studied the S-asymptotically ω-periodic function (see Henríquez *et al.* 2008a, 2008b; Pierri and Nicola, 2009), which is different in ergodicity from the asymptotically ω-periodic function. For more results on S-asymptotic ω-periodicity and its applications to some different differential equations, please refer to Andrade and Cuevas (2010), Andrade *et al.* (2021), Bedi *et al.* (2021), Brindle and N'Guérékata (2019a), Brindle and N'Guérékata (2019b), Brindle and N'Guérékata (2019c), Cuevas and Souza (2010), Cuevas and Lizama (2010), Henríquez *et al.* (2013), Hernández and Pierri (2018), Oueama-Guengai and N'Guérékata (2018), Oueama-Guengai (2018), Pierri (2012), and Shu *et al.* (2015) and references cited therein. The concept of pseudo S-asymptotic ω-periodicity was presented in Pierri and Rolnik (2013), which can be seen as an extension of the aforementioned S-asymptotic ω-periodicity. Moreover, Stepanov-like S-asymptotically ω-periodic function was presented by Henríquez (2013), which was further developed as Stepanov-like weighted pseudo S-asymptotically periodic function by Xia (2015b). The more development and applications of (pseudo) S-asymptotically ω-periodic function can be referred to Cuevas *et al.* (2014), Dimbour (2020), He *et al.* (2020), Henríquez *et al.* (2016), Xia (2015a) and Yang and Wang (2019) and references therein. We can also refer to Alvarez *et al.* (2018), Chaouchi *et al.* (2020), Mophou and N'Guérékata (2020), and Khalladi *et al.* (2021) for a more generalized notion of an ω-periodic function. However, there are few results on pseudo S-asymptotic ω-antiperiodicity corresponding to the ω-antiperiodicity yet. It is also seen from Hasler and N'Guérékata (2014) that a Bloch-type periodic function may not be Bloch-type periodic with a perturbation vanishing at infinity, but it can be asymptotically Bloch-type periodic. Thus, it is meaningful to extend the Bloch-type periodic function such that it can remain certain quasi Bloch-type periodic under differently small perturbations. In this chapter, we introduce some generalized Bloch-type periodic functions and present some fundamental properties on spaces of such functions, most of which can be found in details in Chang and Wei (2021a, 2021b, 2022), and Wei and Chang (2022).

2.1   Bloch-Type Periodic Functions

In this section, we recall the notion of Bloch-type periodic function and some fundamental properties of such functions. For more detailed results and their proof, we can refer to Hasler and N'Guérékata (2014).

Definition 2.1. For given $\omega, k \in \mathbb{R}$, a function $f \in BC(\mathbb{R}, X)$ is said to be Bloch-type periodic (or (ω, k)-Bloch periodic) if for all $t \in \mathbb{R}$,

$$f(t + \omega) = e^{ik\omega} f(t).$$

We will denote by $BP_{\omega, k}(\mathbb{R}, X)$ the space of all (ω, k)-Bloch periodic functions.

Remark 2.1. From Definition 2.1, we can easily have the following assertions:

(1) If f is (ω, k)-Bloch periodic with $k\omega = 2\pi$, then f is simply ω-periodic; if $k\omega = \pi$, then f is ω-antiperiodic.
(2) If $A \in \mathcal{B}(X)$ and f is a Bloch-type periodic X-valued function, then Af is also Bloch-type periodic with the same period ω and constant k.

Example 2.1. Let $u(x, t)$ be a regular solution to the following heat equation:

$$\begin{cases} u_t(x, t) = u_{xx}(x, t), & t > 0, \ x \in \mathbb{R}, \\ u(x, 0) = \phi(x). \end{cases}$$

It is known that

$$u(x, t) = \frac{1}{2\sqrt{\pi t}} \int_{-\infty}^{+\infty} e^{-\frac{(x-s)^2}{4t}} \phi(s) ds.$$

Fix $t^* > 0$ and assume that $\phi(x)$ is a Bloch-type periodic function. Then $u(x, t)$ is also Bloch-type periodic with respect to x.

Indeed, if ϕ is (ω, k)-periodic, then we have

$$u(x + \omega, t^*) = \frac{1}{2\sqrt{\pi t^*}} \int_{-\infty}^{+\infty} e^{-\frac{(x+\omega-s)^2}{4t^*}} \phi(s) ds.$$

By the change of variable $\sigma = s - \omega$, we obtain

$$u(x + \omega, t^*) = \frac{1}{2\sqrt{\pi t^*}} \int_{-\infty}^{+\infty} e^{-\frac{(x-\sigma)^2}{4t^*}} \phi(\sigma + \omega) d\sigma$$

$$= e^{ik\omega} \frac{1}{2\sqrt{\pi t^*}} \int_{-\infty}^{+\infty} e^{-\frac{(x-\sigma)^2}{4t^*}} \phi(\sigma) d\sigma$$

$$= e^{ik\omega} u(x, t^*).$$

Example 2.2. Let f be (ω, k)-Bloch periodic. Then the function

$$F_a(t) := \int_t^{t+a} f(s) ds$$

is also (ω, k)-Bloch periodic. Just apply the same change of variable as Example 2.1.

Example 2.3. Let $u(x,t)$ be a regular solution to the following wave equation:

$$\begin{cases} u_{tt}(x,t) = u_{xx}(x,t), & t > 0, \, x \in \mathbb{R}, \\ u(x,0) = \phi(x), & u_t'(x,0) = \varphi(x). \end{cases}$$

It is known that

$$u(x,t) = \frac{1}{2}[\phi(x+t) + \phi(x-t)] + \frac{1}{2}\int_{x-t}^{x+t} \varphi(s)ds.$$

Fix $t^* > 0$ and assume that $\phi(x), \varphi(x)$ are (ω, k)-Bloch periodic. Then $u(x,t)$ is also (ω, k)-Bloch periodic with respect to x.

Lemma 2.1. *Let* $f_1, f_2, f \in BP_{\omega,k}(\mathbb{R}, X)$. *Then the following results hold:*

(1) $f_1 + f_2 \in BP_{\omega,k}(\mathbb{R}, X)$, *and* $cf \in BP_{\omega,k}(\mathbb{R}, X)$ *for each* $c \in \mathbb{C}$.
(2) *The translated* $f_a(t) := f(t+a) \in BP_{\omega,k}(\mathbb{R}, X)$ *for each* $a \in \mathbb{R}$.
(3) *Let* $f_n \in BP_{\omega,k}(\mathbb{R}, X)$. *If* $f_n \to f$ *uniformly on* $\mathbb{R}$, *then* $f \in BP_{\omega,k}(\mathbb{R}, X)$.
(4) *The space* $BP_{\omega,k}(\mathbb{R}, X)$ *is a Banach space equipped with the sup-norm.*
(5) *Let* $\{\mathcal{S}(t)\}_{t \geq 0} \subseteq \mathcal{B}(X)$ *be a uniformly integrable and strongly continuous family. If* $f \in BP_{\omega,k}(\mathbb{R}, X)$, *then*

$$u(t) := \int_{-\infty}^{t} \mathcal{S}(t-s)f(s)ds \in BP_{\omega,k}(\mathbb{R}, X).$$

Lemma 2.2. *Let* $F \in BC(\mathbb{R} \times X, X)$. *The following properties are equivalent:*

(I) *For each* $\varphi \in BP_{\omega,k}(\mathbb{R}, X)$, $F(\cdot, \varphi(\cdot)) \in BP_{\omega,k}(\mathbb{R}, X)$.
(II) *For every* $(t,x) \in \mathbb{R} \times X$, $F(t+\omega, e^{ik\omega}x) = e^{ik\omega}F(t,x)$.

Lemma 2.3. *Let* $g \in BP_{\omega,k}(\mathbb{R}, X)$ *and* $\epsilon > 0$ *be given. Then there exist* $s_1, \ldots, s_m \in \mathbb{R}$ *such that*

$$\mathbb{R} = \bigcup_{m}^{i=1} (s_i + C_\epsilon),$$

where

$$C_\epsilon := \{t \in \mathbb{R} : \|g(t) - g(0)\| < \epsilon\}.$$

Definition 2.2. A function $f \in BC(\mathbb{R}_+, X)$ is said to be asymptotically Bloch-type periodic (or asymptotically (ω, k)-Bloch periodic) if there exist

$g \in BP_{\omega,k}(\mathbb{R}, X)$ and $h \in C_0(\mathbb{R}_+, X)$ such that

$$f = g + h,$$

where g and h are called, respectively, the principal and corrective terms of f. We denote by $ABP_{\omega,k}(\mathbb{R}, X)$ the space of all asymptotically (ω, k)-Bloch periodic X-valued functions.

Lemma 2.4. *The decomposition of an asymptotically Bloch-type periodic function is unique.*

Remark 2.2. The space $ABP_{\omega,k}(\mathbb{R}, X)$ is a Banach space under the sup-norm, noting that $\{g(t) : t \in \mathbb{R}\} \subseteq \overline{\{f(t) : t \in \mathbb{R}_+\}}$, where g is the principal term of f in Definition 2.2.

2.2 Pseudo Bloch-Type Periodic Functions

This section is mainly concerned with the pseudo Bloch-type periodic function and its basic properties.

In convenience, we introduce the following spaces:

$$\mathscr{E}(\mathbb{R}, X) := \left\{ h \in BC(\mathbb{R}, X) : \lim_{r \to \infty} \frac{1}{2r} \int_{-r}^{r} \|h(t)\| dt = 0 \right\};$$

$$\mathscr{E}(\mathbb{R} \times X, X) := \left\{ h \in BC(\mathbb{R} \times X, X) : \lim_{r \to \infty} \frac{1}{2r} \int_{-r}^{r} \|h(t, x)\| dt = 0 \right.$$

$$\left. \text{uniformly for } x \text{ in any compact subset of } X \right\}.$$

Lemma 2.5 (Blot *et al.*, 2012). *Let $f \in BC(\mathbb{R}, X)$. Then we have the following results:*

(i) *$f \in \mathscr{E}(\mathbb{R}, X)$ if and only if for every $\epsilon > 0$,*

$$\lim_{r \to \infty} \frac{1}{2r} mes(M_{r,\epsilon}(f)) = 0, \text{ where } mes(M_{r,\epsilon}(f)) = \int_{M_{r,\epsilon}(f)} dt.$$

(ii) *The space $\mathscr{E}(\mathbb{R}, X)$ is translation invariant.*
(iii) *The space $\mathscr{E}(\mathbb{R}, X)$ is a Banach space with the sup-norm.*

Definition 2.3. A function $f \in BC(\mathbb{R}, X)$ is said to be pseudo Bloch-type periodic (or pseudo (ω, k)-Bloch periodic), if there exists $g \in BP_{\omega,k}(\mathbb{R}, X)$ and $h \in \mathscr{E}(\mathbb{R}, X)$ such that

$$f = g + h.$$

We denote the set of all such functions by $PBP_{\omega,k}(\mathbb{R}, X)$.

Lemma 2.6. *Let* $f \in PBP_{\omega,k}(\mathbb{R}, X)$ *be such that* $f = g + h$, *where* $g \in BP_{\omega,k}(\mathbb{R}, X)$ *and* $h \in \mathscr{E}(\mathbb{R}, X)$, *then*

$$\{g(t) : t \in \mathbb{R}\} \subseteq \overline{\{f(t) : t \in \mathbb{R}\}}.$$

Proof. Assume that the assertion is not true, then there exists $t_0 \in \mathbb{R}$ such that $g(t_0) \notin \overline{\{f(t) : t \in \mathbb{R}\}}$. Without loss of generality, let $t_0 = 0$, then there exists $\epsilon > 0$, such that

$$\|g(0) - f(t)\| \geq 2\epsilon, \quad \text{for all } t \in \mathbb{R}.$$

It follows that

$$\|h(t)\| = \|f(t) - g(t)\| \geq \|f(t) - g(0)\| - \|g(0) - g(t)\| \geq \epsilon,$$

for all $t \in C_\epsilon$, where C_ϵ is given in Lemma 2.3. Hence,

$$\|h(t - s_i)\| \geq \epsilon,$$

for each $i \in 1, \ldots, m$ and $t \in s_i + C_\epsilon$. Now, we define the function $\mathcal{H}(t)$ by

$$\mathcal{H}(t) = \sum_{i=1}^{m} \|h(t - s_i)\|.$$

From the above inequality, we can see that

$$\mathcal{H}(t) \geq \epsilon,$$

for all $t \in \mathbb{R}$. On the other hand, by the translation invariance of $\mathscr{E}(\mathbb{R}, X)$, we conclude that $h(t - s_i) \in \mathscr{E}(\mathbb{R}, X)$ for all $i \in 1, \ldots, m$, and thus $\mathcal{H} \in \mathscr{E}(\mathbb{R}, X)$, which is contradiction to $\mathcal{H}(t) \geq \epsilon$. So, the conclusion is true. $\square$

Proposition 2.1. *The decomposition of a pseudo Bloch-type periodic function is unique.*

Proof. Assume that $f = g_1 + h_1 = g_2 + h_2$ with $g_i \in BP_{\omega,k}(\mathbb{R}, X)$ and $h_i \in \mathscr{E}(\mathbb{R}, X)$ for $i = 1, 2$. Then we have $0 = (g_1 - g_2) + (h_1 - h_2) \in PBP_{\omega,k}(\mathbb{R}, X)$ with $g_1 - g_2 \in BP_{\omega,k}(\mathbb{R}, X)$ and $h_1 - h_2 \in \mathscr{E}(\mathbb{R}, X)$. From Lemma 2.6, we obtain $(g_1 - g_2)(\mathbb{R}) \subseteq \{0\}$. Hence, we have $g_1 = g_2$ and $h_1 = h_2$. $\square$

Lemma 2.7. *Let $f_1, f_2, f \in PBP_{\omega,k}(\mathbb{R}, X)$. Then the following results are satisfied:*

(1) $f_1 + f_2 \in PBP_{\omega,k}(\mathbb{R}, X)$, *and* $cf \in PBP_{\omega,k}(\mathbb{R}, X)$ *for each* $c \in \mathbb{C}$.
(2) *The translated* $f_a(t) := f(t + a) \in PBP_{\omega,k}(\mathbb{R}, X)$ *for each* $a \in \mathbb{R}$.

Proof. Note that $BP_{\omega,k}(\mathbb{R}, X)$ and $\mathscr{E}(\mathbb{R}, X)$ are Banach space equipped with the sup-norm, they are closed to addition and multiplication. Hence, it is obvious that $f_1 + f_2$ and cf, for any $c \in \mathbb{C}$ are also in $PBP_{\omega,k}(\mathbb{R}, X)$. In view of Lemma 2.1(2) and Lemma 2.5(ii), we can easily deduce that $f_a(t) := f(t + a)$ also belongs to $PBP_{\omega,k}(\mathbb{R}, X)$ for any $a \in \mathbb{R}$. $\square$

Lemma 2.8. *The space $PBP_{\omega,k}(\mathbb{R}, X)$ is a Banach space equipped with the sup-norm.*

Proof. Let $\{f_n\}$ be a Cauchy sequence in $PBP_{\omega,k}(\mathbb{R}, X)$. We can write $f_n = g_n + h_n$ with $g_n \in BP_{\omega,k}(\mathbb{R}, X), h_n \in \mathscr{E}(\mathbb{R}, X)$. From Lemma 2.6, $\|g_n - g_m\|_\infty \leq \|f_n - f_m\|_\infty$, therefore $\{g_n\}$ is a Cauchy sequence in the Banach space $(BP_{\omega,k}(\mathbb{R}, X), \|\cdot\|_\infty)$. So, $\{h_n\} = \{f_n - g_n\}$ is also a Cauchy sequence in the Banach space $(\mathscr{E}(\mathbb{R}, X), \|\cdot\|_\infty)$. Then we can conclude that $\lim_{n \to \infty} g_n = g \in BP_{\omega,k}(\mathbb{R}, X)$ and $\lim_{n \to \infty} h_n = h \in \mathscr{E}(R, X)$, and finally we have $\lim_{n \to \infty} f_n = g + h \in PBP_{\omega,k}(\mathbb{R}, X)$. The above arguments imply that $PBP_{\omega,k}(\mathbb{R}, X)$ is a closed subspace of $BC(\mathbb{R}, X)$, and thus $PBP_{\omega,k}(\mathbb{R}, X)$ is also a Banach space equipped with the sup-norm. $\square$

Lemma 2.9. *Let $\{\mathcal{S}(t)\}_{t \geq 0} \subseteq \mathcal{B}(X)$ be a uniformly integrable and strongly continuous family, and $h \in \mathscr{E}(\mathbb{R}, X)$, then*

$$H_1(t) := \int_{-\infty}^{t} \mathcal{S}(t - s)h(s)ds \in \mathscr{E}(\mathbb{R}, X).$$

Proof. Since the operator family $\{\mathcal{S}(t)\}_{t \geq 0} \subseteq \mathcal{B}(X)$ is assumed to be uniformly integrable, we have

$$\int_{0}^{\infty} \|\mathcal{S}(t)\|dt < \infty.$$

By the Fubini theorem, we have

$$\frac{1}{2r}\int_{-r}^{r}\|H_1(t)\|dt \leq \frac{1}{2r}\int_{-r}^{r}\left[\int_{-\infty}^{t}\|\mathcal{S}(t-s)\|\|h(s)\|ds\right]dt$$

$$\leq \frac{1}{2r}\int_{-r}^{r}\left[\int_{0}^{\infty}\|\mathcal{S}(s)\|\|h(t-s)\|ds\right]dt$$

$$= \int_{0}^{\infty}\|\mathcal{S}(s)\|\left[\frac{1}{2r}\int_{-r}^{r}\|h(t-s)\|dt\right]ds.$$

Owing to $h \in \mathscr{E}(\mathbb{R}, X)$, it is deduced that

$$\frac{1}{2r}\int_{-r}^{r}\|H_1(t)\|dt \to 0, \text{ as } r \to \infty$$

by Lemma 2.5(ii) and Lebesgue dominated convergence theorem. □

Theorem 2.1. *Let $\{\mathcal{S}(t)\}_{t\geq 0} \subseteq \mathcal{B}(X)$ be a uniformly integrable and strongly continuous family. If $f \in PBP_{\omega,k}(\mathbb{R}, X)$, then*

$$u(t) := \int_{-\infty}^{t}\mathcal{S}(t-s)f(s)ds \in PBP_{\omega,k}(\mathbb{R}, X).$$

Proof. Let $f \in PBP_{\omega,k}(\mathbb{R}, X)$ be such that $f(t) = g(t) + h(t)$ with $g \in BP_{\omega,k}(\mathbb{R}, X)$ and $h \in \mathscr{E}(\mathbb{R}, X)$, then

$$\int_{-\infty}^{t}\mathcal{S}(t-s)f(s)ds = \int_{-\infty}^{t}\mathcal{S}(t-s)g(s)ds + \int_{-\infty}^{t}\mathcal{S}(t-s)h(s)ds$$

$$:= u_1(t) + H_1(t).$$

In view of Lemma 2.1(5) and Lemma 2.9, we have $u_1(t) \in BP_{\omega,k}(\mathbb{R}, X)$ and $H_1(t) \in \mathscr{E}(\mathbb{R}, X)$, respectively. Thus, we deduce that

$$\int_{-\infty}^{t}\mathcal{S}(t-s)f(s)ds \in PBP_{\omega,k}(\mathbb{R}, X).$$

The proof is ended. □

2.3   Weighted Pseudo Bloch-Type Periodic Functions

In this section, we introduce the notion of weighted pseudo Bloch-type periodic functions and some basic properties.

Let $\mathcal{U}$ denote the set of all functions $\rho : \mathbb{R} \to (0, \infty)$, which are locally integrable over $\mathbb{R}$ such that $\rho > 0$ almost everywhere. For a given $r > 0$ and for each $\rho \in \mathcal{U}$, we set

$$\mu(r, \rho) = \int_{-r}^{r} \rho(t)dt.$$

Define the set

$$\mathcal{U}_\infty = \Big\{ \rho \in \mathcal{U} : \lim_{r \to \infty} \mu(r, \rho) = \infty \Big\}.$$

For $\rho \in \mathcal{U}_\infty$, we define spaces $\mathscr{E}(\mathbb{R}, X, \rho)$, $\mathscr{E}(\mathbb{R} \times X, X, \rho)$, respectively, by

$$\mathscr{E}(\mathbb{R}, X, \rho) := \Big\{ h \in BC(\mathbb{R}, X) : \lim_{r \to \infty} \frac{1}{\mu(r, \rho)} \int_{-r}^{r} \|h(t)\|\rho(t)dt = 0 \Big\};$$

$$\mathscr{E}(\mathbb{R} \times X, X, \rho) := \Big\{ h \in BC(\mathbb{R} \times X, X) : \lim_{r \to \infty} \frac{1}{\mu(r, \rho)} \int_{-r}^{r} \|h(t, x)\|\rho(t)dt$$

$$= 0 \text{ uniformly for } x \text{ in any compact subset of X} \Big\}.$$

Furthermore, for $\rho \in \mathcal{U}_\infty$, let us list the following hypothesis:

$$\mathrm{H}_\rho : \text{ For all } \tau \in \mathbb{R}, \ \limsup_{|t| \to +\infty} \frac{\rho(t + \tau)}{\rho(t)} < +\infty.$$

In convenience, we define the set

$$\mathcal{U}_{\mathrm{inv}} := \{ \rho \in \mathcal{U}_\infty : \rho \text{ satisfies the hypothesis } \mathrm{H}_\rho \}.$$

Remark 2.3 (Blot *et al.*, 2012). The hypothesis H_ρ implies that for all $\tau > 0$,

$$\limsup_{r \to \infty} \frac{\mu(r + \tau, \rho)}{\mu(r, \rho)} < +\infty.$$

Lemma 2.10 (Blot *et al.*, 2012). *Let $f \in BC(\mathbb{R}, X)$ and $\rho \in \mathcal{U}_\infty$. Then we have the following results:*

(a) *$f \in \mathscr{E}(\mathbb{R}, X, \rho)$ if and only if for every $\epsilon > 0$,*

$$\lim_{r \to \infty} \frac{1}{\mu(r, \rho)} \int_{M_{r,\epsilon}(f)} \rho(t)dt = 0,$$

where $M_{r,\epsilon}(f) = \{ t \in [-r, r] : \|f(t)\| \geq \epsilon \}$.

(b) *If ρ satisfies the hypothesis H_ρ, then the space $\mathscr{E}(\mathbb{R}, X, \rho)$ is translation invariant, i.e., for each $\tau \in \mathbb{R}$, $f \in \mathscr{E}(\mathbb{R}, X, \rho)$ implies that $f_\tau \in \mathscr{E}(\mathbb{R}, X, \rho)$, where $f_\tau(t) = f(t + \tau)$, $t \in \mathbb{R}$.*

(c) *The space $\mathscr{E}(\mathbb{R}, X, \rho)$ is a Banach space equipped with the sup-norm.*

Definition 2.4. Let $\rho \in \mathcal{U}_\infty$. A function $f \in BC(\mathbb{R}, X)$ is said to be weighted pseudo Bloch-type periodic (or weighted pseudo (ω, k)-Bloch periodic), if there exists $g \in BP_{\omega,k}(\mathbb{R}, X)$ and $h \in \mathscr{E}(\mathbb{R}, X, \rho)$ such that

$$f = g + h.$$

The set of all such functions is denoted by $WPBP_{\omega,k}(\mathbb{R}, X, \rho)$.

Lemma 2.11. *Let $\rho \in \mathcal{U}_{\mathrm{inv}}$ and $f \in WPBP_{\omega,k}(\mathbb{R}, X, \rho)$ be such that $f = g + h$ with $g \in BP_{\omega,k}(\mathbb{R}, X)$ and $h \in \mathscr{E}(\mathbb{R}, X, \rho)$. Then*

$$\{g(t) : t \in \mathbb{R}\} \subseteq \overline{\{f(t) : t \in \mathbb{R}\}}.$$

Proof. The proof is similar to that of Lemma 2.6, and by the translation invariance of $\mathscr{E}(\mathbb{R}, X, \rho)$ owing to the hypothesis H_ρ, we can conclude that the assertion is true. □

Proposition 2.2. *Let $\rho \in \mathcal{U}_{\mathrm{inv}}$. Then the decomposition of a weighted pseudo Bloch-type periodic function is unique.*

Proof. Taking into account Lemma 2.11, the conclusion can be similarly conducted as that of Proposition 2.1. □

Lemma 2.12. *The following results are satisfied:*

(1) *Let $\rho \in \mathcal{U}_\infty$, and $f_1, f_2, f \in WPBP_{\omega,k}(\mathbb{R}, X, \rho)$, then $f_1 + f_2 \in WPBP_{\omega,k}(\mathbb{R}, X)$, and $cf \in WPBP_{\omega,k}(\mathbb{R}, X)$ for each $c \in \mathbb{C}$.*
(2) *Let $\rho \in \mathcal{U}_{\mathrm{inv}}$, then $f_a(t) := f(t+a) \in WPBP_{\omega,k}(\mathbb{R}, X, \rho)$ for any $a \in \mathbb{R}$.*

Proof. Considering $BP_{\omega,k}(\mathbb{R}, X)$ and $\mathscr{E}(\mathbb{R}, X, \rho)$ are Banach spaces equipped with the sup-norm, and closed to addition and multiplication, thus $f_1 + f_2$ and cf are also in $WPBP_{\omega,k}(\mathbb{R}, X, \rho)$ for any $c \in \mathbb{C}$. For $\rho \in \mathcal{U}_{inv}$, the space $\mathscr{E}(\mathbb{R}, X, \rho)$ is translation invariant by Lemma 2.10(b), and thus $f_a(t) := f(t + a) \in WPBP_{\omega,k}(\mathbb{R}, X, \rho)$ by Lemma 2.1(2). □

Lemma 2.13. *Let $\rho \in \mathcal{U}_{\mathrm{inv}}$. Then the space $WPBP_{\omega,k}(\mathbb{R}, X, \rho)$ is a Banach space equipped with the sup-norm.*

Proof. The proof can be conducted similarly to that of Lemma 2.8 by taking into account Lemma 2.11 and Proposition 2.2. □

Taking into account Lemma 2.10(b), we can conduct similarly to the proof of Lemma 2.9 and obtain the following result.

Lemma 2.14. *Let $\{S(t)\}_{t\geq 0} \subseteq \mathcal{B}(X)$ be a uniformly integrable and strongly continuous family, and $h \in \mathcal{E}(\mathbb{R}, X, \rho)$ with $\rho \in \mathcal{U}_{\mathrm{inv}}$, then*

$$H_2(t) := \int_{-\infty}^{t} S(t-s)h(s)ds \in \mathcal{E}(\mathbb{R}, X, \rho).$$

Theorem 2.2. *Let $\rho \in \mathcal{U}_{\mathrm{inv}}$ and $\{S(t)\}_{t\geq 0} \subseteq \mathcal{B}(X)$ be a uniformly integrable and strongly continuous family. If $f \in WPBP_{\omega,k}(\mathbb{R}, X, \rho)$, then*

$$u(t) := \int_{-\infty}^{t} S(t-s)f(s)ds \in WPBP_{\omega,k}(\mathbb{R}, X, \rho).$$

Proof. The proof can be conducted similarly as Theorem 2.1 by Lemma 2.1(5) and Lemma 2.14. □

We shall prove some composition results on (weighted) pseudo Bloch-type periodic functions in the following. We introduce the following sets defined, respectively, by

$$\mathscr{S}(r,\rho) = \left\{\nu : \mathbb{R} \to \mathbb{R}_+ : \lim_{r\to\infty} \frac{1}{\mu(r,\rho)} \int_{-r}^{r} \nu(t)\rho(t)dt < \infty \right\}, \quad \rho \in \mathcal{U}_\infty,$$

and

$$\mathscr{S}(r) = \left\{\nu : \mathbb{R} \to \mathbb{R}_+ : \lim_{r\to\infty} \frac{1}{2r} \int_{-r}^{r} \nu(t)dt < \infty \right\}.$$

We also assume that the following condition hold:

(A1) $g \in BC(\mathbb{R} \times X, X)$, and

$$g(t + \omega, e^{ik\omega}x) = e^{ik\omega}g(t, x)$$

for all $(t, x) \in \mathbb{R} \times X$.

With the set $\mathscr{S}(r, \rho)$ (resp. $\mathscr{S}(r)$) given above, we first establish two basic composition theorems (see Theorems 2.3–2.4). And then, some corollaries are attained according to Theorems 2.3–2.4, some of which are even new for (weighted) pseudo antiperiodic functions.

Theorem 2.3. *Let $\rho \in \mathcal{U}_{\mathrm{inv}}$, $f = g + h \in BC(\mathbb{R} \times X, X)$ with g satisfying the condition (A1) and $h \in \mathcal{E}(\mathbb{R} \times X, X, \rho)$. Assume that the following condition holds:*

(A2) *There exists a function $\mathcal{L}(\cdot) \in \mathscr{S}(r, \rho)$ such that for all x, $y \in X$ and $t \in \mathbb{R}$,*

$$\|f(t, x) - f(t, y)\| \leq \mathcal{L}(t)\|x - y\|.$$

Then for each $x(\cdot) \in WPBP_{\omega,k}(\mathbb{R}, X, \rho)$, we have

$$f(\cdot, x(\cdot)) \in WPBP_{\omega,k}(\mathbb{R}, X, \rho).$$

Proof. Let $x(t) := \alpha(t) + \beta(t) \in WPBP_{\omega,k}(\mathbb{R}, X, \rho)$, where $\alpha(t) \in BP_{\omega,k}(\mathbb{R}, X)$ and $\beta(t) \in \mathscr{E}(\mathbb{R}, X, \rho)$. The function f can be rewritten as

$$f(t, x(t)) = g(t, \alpha(t)) + f(t, x(t)) - g(t, \alpha(t))$$

$$= g(t, \alpha(t)) + f(t, x(t)) - f(t, \alpha(t)) + h(t, \alpha(t)).$$

Define

$$G(t) := f(t, x(t)) - f(t, \alpha(t)), \quad H(t) := h(t, \alpha(t)).$$

According to (A1) and Lemma 2.2, we have $g(t, \alpha(t)) \in BP_{\omega,k}(\mathbb{R}, X)$. To complete the proof, it is enough to show that $G(t) \in \mathscr{E}(\mathbb{R}, X, \rho)$ and $H(t) \in \mathscr{E}(\mathbb{R}, X, \rho)$.

We firstly show that $G \in \mathscr{E}(\mathbb{R}, X, \rho)$. Clearly, $f(t, x(t)) - f(t, \alpha(t)) \in BC(\mathbb{R}, X)$, without loss of generality, we assume that $\|f(t, x(t)) - f(t, \alpha(t))\| \leq C$ for some positive constant C. By the fact that $\beta \in \mathscr{E}(\mathbb{R}, X, \rho)$ and Lemma 2.10, for any sufficiently small $\epsilon > 0$, we have

$$\lim_{r \to \infty} \frac{1}{\mu(r, \rho)} \int_{M_{r,\epsilon}(\beta)} \rho(t)dt = 0,$$

where $M_{r,\epsilon}(\beta) = \{t \in [-r, r] : \|\beta(t)\| \geq \epsilon\}$. Therefore,

$$\frac{1}{\mu(r, \rho)} \int_{[-r,r]} \|G(t)\|\rho(t)dt$$

$$= \frac{1}{\mu(r, \rho)} \int_{[-r,r]} \|f(t, x(t)) - f(t, \alpha(t))\|\rho(t)dt$$

$$= \frac{1}{\mu(r, \rho)} \int_{M_{r,\epsilon}(\beta)} \|f(t, x(t)) - f(t, \alpha(t))\|\rho(t)dt$$

$$+ \frac{1}{\mu(r, \rho)} \int_{[-r,r] \backslash M_{r,\epsilon}(\beta)} \|f(t, x(t)) - f(t, \alpha(t))\|\rho(t)dt$$

$$\leq C \frac{1}{\mu(r, \rho)} \int_{M_{r,\epsilon}(\beta)} \rho(t)dt + \frac{1}{\mu(r, \rho)} \int_{[-r,r] \backslash M_{r,\epsilon}(\beta)} \mathcal{L}(t)\|\beta(t)\|\rho(t)dt$$

$$\leq C \frac{1}{\mu(r, \rho)} \int_{M_{r,\epsilon}(\beta)} \rho(t)dt + \epsilon \frac{1}{\mu(r, \rho)} \int_{[-r,r]} \mathcal{L}(t)\rho(t)dt.$$

Taking into account that $\mathcal{L}(\cdot) \in \mathscr{S}(r, \rho)$ and ϵ sufficiently small, we obtain

$$\lim_{r \to \infty} \frac{1}{\mu(r, \rho)} \int_{[-r,r]} \|G(t)\| \rho(t) dt = 0,$$

which shows that $G(\cdot) \in \mathscr{E}(\mathbb{R}, X, \rho)$.

In the following, we show that $H(t) \in \mathscr{E}(\mathbb{R}, X, \rho)$. By the continuity of α, we deduce that $\mathfrak{I} := \alpha([-r, r])$ is a compact set. Note that $H(t) := h(t, \alpha(t))$ is continuous on a compact interval $[-r, r]$, so it is uniformly continuous on $[-r, r]$. Thus, there exist finite open balls $\mathfrak{B}_k$ $(k = 1, 2, \ldots, m)$ with center $x_k \in \mathfrak{I}$ and radius $\delta > 0$ small enough such that

$$\mathfrak{I} \subseteq \bigcup_{k=1}^{m} \mathfrak{B}_k$$

and

$$\|h(t, \alpha(t)) - h(t, x_k)\| < \frac{\epsilon}{2},$$

for any sufficiently small $\epsilon > 0$ with $\alpha(t) \in \mathfrak{B}_k, t \in [-r, r]$. Define the open set $\mathcal{O}_k = \{t \in [-r, r] : \alpha(t) \in \mathfrak{B}_k\}$ such that $[-r, r] = \bigcup_{k=1}^{m} \mathcal{O}_k$. Let

$$\mathfrak{E}_1 = \mathcal{O}_1, \mathfrak{E}_k = \mathcal{O}_k \backslash \bigcup_{j=1}^{k-1} \mathcal{O}_j, \quad 2 \le k \le m,$$

then $\mathfrak{E}_i \cap \mathfrak{E}_j = \emptyset$ if $i \ne j$, $1 \le i, j \le m$. Thus,

$$\{t \in [-r, r] : \|h(t, \alpha(t))\| \ge \epsilon\}$$

$$\subseteq \bigcup_{k=1}^{m} \{t \in \mathfrak{E}_k : \|h(t, \alpha(t)) - h(t, x_k)\| + \|h(t, x_k)\| \ge \epsilon\}$$

$$\subseteq \bigcup_{k=1}^{m} \left(\left\{ t \in \mathfrak{E}_k : \|h(t, \alpha(t)) - h(t, x_k)\| \ge \frac{\epsilon}{2} \right\} \right.$$

$$\left. \times \bigcup \left\{ t \in \mathfrak{E}_k : \|h(t, x_k)\| \ge \frac{\epsilon}{2} \right\} \right).$$

Noticing that $\alpha(t) \in \mathfrak{B}_k$, so $\left\{ t \in \mathfrak{E}_k : \|h(t, \alpha(t)) - h(t, x_k)\| \ge \frac{\epsilon}{2} \right\} = \emptyset$ for $1 \le k \le m$. Hence,

$$\frac{1}{\mu(r, \rho)} \int_{M_{r,\epsilon}(h(t,\alpha(t)))} \rho(t) dt \le \sum_{k=1}^{m} \frac{1}{\mu(r, \rho)} \int_{M_{r,\frac{\epsilon}{2}}(h(t,x_k))} \rho(t) dt.$$

Taking into account that $h(t, x) \in \mathscr{E}(\mathbb{R} \times X, X, \rho)$, we obtain

$$\frac{1}{\mu(r, \rho)} \int_{M_{r,\frac{\epsilon}{2}}(h(t,x_k))} \rho(t) dt \to 0, \quad \text{as } r \to \infty,$$

which implies that

$$\lim_{r \to \infty} \frac{1}{\mu(r,\rho)} \int_{M_{r,\epsilon}(h(t,\alpha(t)))} \rho(t)dt = 0.$$

Therefore, $H(t) := h(t,\alpha(t)) \in \mathscr{E}(\mathbb{R}, X, \rho)$. This completes the proof. $\square$

According to Lemma 2.5 and Theorem 2.3, we can easily obtain the following result for pseudo Bloch-type periodic functions.

Corollary 2.1. *Let* $f = g + h \in BC(\mathbb{R} \times X, X)$ *with* $g \in BC(\mathbb{R} \times X, X)$ *and* $h \in \mathscr{E}(\mathbb{R} \times X, X)$. *Assume that the condition (A1) and the following assumption:*

(A3) *There exists a function* $\mathcal{L}(\cdot) \in \mathscr{S}(r)$ *such that for all* $x, y \in X$ *and* $t \in \mathbb{R}$,

$$\|f(t,x) - f(t,y)\| \leq \mathcal{L}(t)\|x - y\|$$

are satisfied. Then for each $x(\cdot) \in PBP_{\omega,k}(\mathbb{R}, X)$, *we have*

$$f(\cdot, x(\cdot)) \in PBP_{\omega,k}(\mathbb{R}, X).$$

We can easily deduce the following corollaries from Theorem 2.3 and Corollary 2.1.

Corollary 2.2 (Lipschitz growth condition). *Assume that* $f \in BC(\mathbb{R} \times X, X)$ *and there exists a constant* $l > 0$ *such that for all* $x, y \in X$ *and* $t \in \mathbb{R}$,

$$\|f(t,x) - f(t,y)\| \leq l\|x - y\|.$$

Then the following conclusions hold:

(1) *If* $f = g + h$ *with* g *satisfying the condition (A1) and* $h \in \mathscr{E}(\mathbb{R} \times X, X)$, *then for each* $x(\cdot) \in PBP_{\omega,k}(\mathbb{R}, X)$, $f(\cdot, x(\cdot)) \in PBP_{\omega,k}(\mathbb{R}, X)$.
(2) *If* $f = g + h$ *with* g *satisfying the condition (A1) and* $h \in \mathscr{E}(\mathbb{R} \times X, X, \rho)$ *for* $\rho \in \mathcal{U}_{\text{inv}}$, *then for each* $x(\cdot) \in WPBP_{\omega,k}(\mathbb{R}, X, \rho)$, $f(\cdot, x(\cdot)) \in WPBP_{\omega,k}(\mathbb{R}, X, \rho)$.

Proof. Note that any positive constant l must belong to the set $\mathscr{S}(r)$ or $\mathscr{S}(r, \rho)$, above conclusions can be deduced directly via Corollary 2.1 and Theorem 2.3. $\square$

Corollary 2.3. *Let* $f = g + h \in BC(\mathbb{R} \times X, X)$ *with* g *satisfying the condition (A1) and* $h \in \mathscr{E}(\mathbb{R} \times X, X)$. *Suppose further that the following condition holds:*

(C1) *There exists a function $L(\cdot) \in L^p(\mathbb{R}, \mathbb{R}_+)(1 \leq p < +\infty)$ such that for all x, $y \in X$ and $t \in \mathbb{R}$,*

$$\|f(t, x) - f(t, y)\| \leq L(t)\|x - y\|.$$

Then for each $x(\cdot) \in PBP_{\omega,k}(\mathbb{R}, X)$, $f(\cdot, x(\cdot)) \in PBP_{\omega,k}(\mathbb{R}, X)$.

Proof. It is only to show that the function $L(t) \in L^p(\mathbb{R}, \mathbb{R}_+)(1 \leq p < +\infty)$ also belongs to the set $\mathscr{S}(r)$. In fact, if $L(t) \in L^1(\mathbb{R}, \mathbb{R}_+)$, it is obvious that $L(t) \in \mathscr{S}(r)$. Let $L(t) \in L^p(\mathbb{R}, \mathbb{R}_+)$ with $1 < p < \infty$, then we have

$$\frac{1}{2r} \int_{-r}^{r} L(t)dt \leq \frac{\left(\int_{-r}^{r} L^p(t)dt\right)^{\frac{1}{p}}}{(2r)^{\frac{1}{p}}} \leq \frac{\left(\int_{-\infty}^{\infty} L^p(t)dt\right)^{\frac{1}{p}}}{(2r)^{\frac{1}{p}}} \to 0, \ r \to \infty,$$

which also implies that $L(t) \in \mathscr{S}(r)$. □

Corollary 2.4. *Let $f = g + h \in BC(\mathbb{R} \times X, X)$ with g satisfying the condition (A1) and $h \in \mathscr{E}(\mathbb{R} \times X, X, \rho)$ for $\rho \in \mathcal{U}_{\text{inv}}$. Suppose further that the following condition holds:*

(C2) *There exists a function $L(\cdot) : \mathbb{R} \to \mathbb{R}_+$ satisfying $L(t)\rho(t) \in L^1(\mathbb{R}, \mathbb{R}_+)$, such that for all x, $y \in X$ and $t \in \mathbb{R}$,*

$$\|f(t, x) - f(t, y)\| \leq L(t)\|x - y\|.$$

Then for each $x(\cdot) \in WPBP_{\omega,k}(\mathbb{R}, X, \rho)$, we have

$$f(\cdot, x(\cdot)) \in WPBP_{\omega,k}(\mathbb{R}, X, \rho).$$

Proof. It is easy to see that if $L(t)\rho(t) \in L^1(\mathbb{R}, \mathbb{R}_+)$, then $L(t) \in \mathscr{S}(r, \rho)$. Thus, the conclusion is a direct consequence of Theorem 2.3. □

Theorem 2.4. *Let $\rho \in \mathcal{U}_{\text{inv}}$, $f = g + h \in BC(\mathbb{R} \times X, X)$ with $g \in BC(\mathbb{R} \times X, X)$ and $h \in \mathscr{E}(\mathbb{R} \times X, X, \rho)$. Assume that the condition (A1) and the following hypothesis hold:*

(A4) *There exists a function $\mathbb{L}(\cdot) \in \mathscr{S}(r, \rho)$ such that for each $\epsilon > 0$, there is a constant $\delta > 0$ satisfying*

$$\|f(t, x) - f(t, y)\| \leq \mathbb{L}(t)\epsilon$$

for all x, y belonging to any bounded subset $\mathcal{Q} \subseteq X$ with $\|x - y\| \leq \delta$ and $t \in \mathbb{R}$.

Then for each $x(\cdot) \in WPBP_{\omega,k}(\mathbb{R}, X, \rho)$, we have

$$f(\cdot, x(\cdot)) \in WPBP_{\omega,k}(\mathbb{R}, X, \rho).$$

Proof. For each $x(t) = \alpha(t) + \beta(t) \in WPBP_{\omega,k}(\mathbb{R}, X, \rho)$ with $\alpha(t) \in BP_{\omega,k}(\mathbb{R}, X)$ and $\beta(t) \in \mathscr{E}(\mathbb{R}, X, \rho)$, we can rewrite the function f in the form:

$$f(t, x(t)) = g(t, \alpha(t)) + f(t, x(t)) - g(t, \alpha(t))$$
$$= g(t, \alpha(t)) + f(t, x(t)) - f(t, \alpha(t)) + h(t, \alpha(t)).$$

Define

$$\Phi(t) := f(t, x(t)) - f(t, \alpha(t)), \ H(t) := h(t, \alpha(t)).$$

By the condition (A1) and Lemma 2.2, we conclude that $g(t, \alpha(t)) \in BP_{\omega,k}(\mathbb{R}, X)$. From the proof of Theorem 2.3, we can also have $H(t) \in \mathscr{E}(\mathbb{R}, X, \rho)$. The remainder is only to show that $\Phi(t) \in \mathscr{E}(\mathbb{R}, X, \rho)$. In fact, $f(t, x(t)) - f(t, \alpha(t)) \in BC(\mathbb{R}, X)$, and we can assume that $\|\Phi(t)\| \leq C$. For $\Phi \in \mathscr{E}(\mathbb{R}, X, \rho)$, it is enough to show that

$$\lim_{r \to \infty} \frac{1}{\mu(r, \rho)} \int_{[-r,r]} \|\Phi(t)\| \rho(t) dt = 0.$$

By Lemma 2.11, $\alpha(\mathbb{R}) \subseteq \overline{x(\mathbb{R})}$ is a bounded set. By the condition (A4) with $Q = \overline{x(\mathbb{R})}$, we see that for each $\epsilon > 0$, there exists a constant $\delta > 0$ such that for all $t \in \mathbb{R}$,

$$\|x - \alpha\| \leq \delta \Rightarrow \|f(t, x(t)) - f(t, \alpha(t))\| \leq \mathbb{L}(t)\epsilon.$$

Considering $M_{r,\epsilon}(\beta) = \{t \in [-r, r] : \|\beta(t)\| > \epsilon\}$, we have

$$M_{r,\mathbb{L}(t)\epsilon}(\Phi) = M_{r,\mathbb{L}(t)\epsilon}(f(t, x(t)) - f(t, \alpha(t)))$$
$$\subseteq M_{r,\delta}(x(t) - \alpha(t)) = M_{r,\delta}(\beta).$$

Hence,

$$\frac{\int_{M_{r,\mathbb{L}(t)\epsilon}(\Phi)} \rho(t) dt}{\mu(r, \rho)} \leq \frac{\int_{M_{r,\delta}(\beta)} \rho(t) dt}{\mu(r, \rho)}.$$

In view of $\beta \in \mathscr{E}(\mathbb{R}, X, \rho)$ and Lemma 2.10, for the above mentioned δ, we have

$$\lim_{r \to \infty} \frac{\int_{M_{r,\delta}(\beta)} \rho(t) dt}{\mu(r, \rho)} = 0,$$

and

$$\lim_{r \to \infty} \frac{\int_{M_{r,\mathbb{L}(t)\epsilon}(\Phi)} \rho(t)dt}{\mu(r,\rho)} = 0. \tag{2.1}$$

Therefore,

$$\frac{1}{\mu(r,\rho)} \int_{[-r,r]} \|\Phi(t)\|\rho(t)dt$$

$$= \frac{1}{\mu(r,\rho)} \int_{[-r,r]\setminus M_{r,\mathbb{L}(t)\epsilon}(\Phi)} \|\Phi(t)\|\rho(t)dt$$

$$+ \frac{1}{\mu(r,\rho)} \int_{M_{r,\mathbb{L}(t)\epsilon}(\Phi)} \|\Phi(t)\|\rho(t)dt$$

$$\leq c \frac{\int_{M_{r,\mathbb{L}(t)\epsilon}(\Phi)} \rho(t)dt}{\mu(r,\rho)} + \epsilon \frac{1}{\mu(r,\rho)} \int_{[-r,r]} \mathbb{L}(t)\rho(t)dt.$$

From (2.1) and the fact $\mathbb{L}(\cdot) \in \mathscr{S}(r,\rho)$, we can see that for any sufficiently small $\epsilon > 0$,

$$\lim_{r \to \infty} \frac{1}{\mu(r,\rho)} \int_{[-r,r]} \|\Phi(t)\|\rho(t)dt = 0,$$

which implies $\Phi(t) \in \mathscr{E}(\mathbb{R}, X, \rho)$. □

From Lemmas 2.5, 2.6 and the main proofs of Theorem 2.4, Corollaries 2.3–2.4, we can also deduce the following results.

Corollary 2.5. *Let $f = g + h \in BC(\mathbb{R} \times X, X)$ with $g \in BC(\mathbb{R} \times X, X)$ and $h \in \mathscr{E}(\mathbb{R} \times X, X)$. Assume that the condition (A1) and the following hypothesis hold:*

(A5) *There exists a function $\mathbb{L}(\cdot) \in \mathscr{S}(r)$ such that for each $\epsilon > 0$, there is a constant $\delta > 0$ satisfying*

$$\|f(t,x) - f(t,y)\| \leq \mathbb{L}(t)\epsilon$$

for all x, y belonging to any bounded subset $\mathcal{Q} \subseteq X$ with $\|x - y\| \leq \delta$ and $t \in \mathbb{R}$.

Then for each $x(\cdot) \in PBP_{\omega,k}(\mathbb{R}, X)$, $f(\cdot, x(\cdot)) \in PBP_{\omega,k}(\mathbb{R}, X)$.

Corollary 2.6. *Let $f = g + h \in BC(\mathbb{R} \times X, X)$ with $g \in BC(\mathbb{R} \times X, X)$ and $h \in \mathscr{E}(\mathbb{R} \times X, X)$. Assume that the condition (A1) and the following hypothesis hold:*

(C3) *There exists a function $L(\cdot) \in L^p(\mathbb{R}, \mathbb{R}_+)(1 \leq p < +\infty)$ such that for each $\epsilon > 0$, there is a constant $\delta > 0$ satisfying*

$$\|f(t, x) - f(t, y)\| \leq L(t)\epsilon$$

for all x, y belonging to any bounded subset $\mathcal{Q} \subseteq X$ with $\|x - y\| \leq \delta$ and $t \in \mathbb{R}$.

Then for each $x(\cdot) \in PBP_{\omega,k}(\mathbb{R}, X)$, $f(\cdot, x(\cdot)) \in PBP_{\omega,k}(\mathbb{R}, X)$.

Corollary 2.7. *Let $\rho \in \mathcal{U}_{\mathrm{inv}}$, $f = g + h \in BC(\mathbb{R} \times X, X)$ with $g \in BC(\mathbb{R} \times X, X)$ and $h \in \mathscr{E}(\mathbb{R} \times X, X, \rho)$. Assume that the condition (A1) and the following hypothesis hold:*

(C4) *There exists a function $L(\cdot) : \mathbb{R} \to \mathbb{R}_+$ satisfying $L(t)\rho(t) \in L^1(\mathbb{R}, \mathbb{R}_+)$, such that for each $\epsilon > 0$, there is a constant $\delta > 0$ satisfying*

$$\|f(t, x) - f(t, y)\| \leq L(t)\epsilon$$

for all x, y belonging to any bounded subset $\mathcal{Q} \subseteq X$ with $\|x - y\| \leq \delta$ and $t \in \mathbb{R}$.

Then for each $x(\cdot) \in WPBP_{\omega,k}(\mathbb{R}, X, \rho)$, we have

$$f(\cdot, x(\cdot)) \in WPBP_{\omega,k}(\mathbb{R}, X, \rho).$$

Corollary 2.8 (Uniformly continuous condition). *Let $f \in BC(\mathbb{R} \times X, X)$ be such that, for each $\epsilon > 0$, there is a constant $\delta > 0$ satisfying*

$$\|f(t, x) - f(t, y)\| \leq \epsilon$$

for all x, y belonging to any bounded subset $\mathcal{Q} \subseteq X$ with $\|x - y\| \leq \delta$ and $t \in \mathbb{R}$. Then the following assertions hold:

(1) *If $f = g + h$ with g satisfying the condition (A1) and $h \in \mathscr{E}(\mathbb{R} \times X, X)$, then for each $x(\cdot) \in PBP_{\omega,k}(\mathbb{R}, X)$, $f(\cdot, x(\cdot)) \in PBP_{\omega,k}(\mathbb{R}, X)$.*

(2) *If $f = g + h$ with g satisfying the condition (A1) and $h \in \mathscr{E}(\mathbb{R} \times X, X, \rho)$ for $\rho \in \mathcal{U}_{\mathrm{inv}}$, then for each $x(\cdot) \in WPBP_{\omega,k}(\mathbb{R}, X, \rho)$, $f(\cdot, x(\cdot)) \in WPBP_{\omega,k}(\mathbb{R}, X, \rho)$.*

Corollary 2.9 (Hölder type growth condition). *Let $\rho \in \mathcal{U}_{\text{inv}}$, $f = g + h \in BC(\mathbb{R} \times X, X)$ with $g \in BC(\mathbb{R} \times X, X)$ and $h \in \mathcal{E}(\mathbb{R} \times X, X, \rho)$ (resp. $h \in \mathcal{E}(\mathbb{R} \times X, X)$). Assume that the condition $(A1)$ and the following hypothesis hold:*

(C5) *There exists a function $\mathbb{L}(\cdot) \in \mathscr{S}(r, \rho)$ (resp. $\mathbb{L}(\cdot) \in \mathscr{S}(r)$) such that for any bounded subset $\mathcal{Q} \subseteq X$*

$$\|f(t, x) - f(t, y)\| \leq \mathbb{L}(t)\|x - y\|^{\vartheta}, \ \vartheta \in (0, 1),$$

for all $x, y \in \mathcal{Q} \subseteq X$ and $t \in \mathbb{R}$.

If $x(\cdot) \in WPBP_{\omega,k}(\mathbb{R}, X, \rho)$ (resp. $x(\cdot) \in PBP_{\omega,k}(\mathbb{R}, X)$), then $f(\cdot, x(\cdot)) \in WPBP_{\omega,k}(\mathbb{R}, X, \rho)$ (respectively, $f(\cdot, x(\cdot)) \in PBP_{\omega,k}(\mathbb{R}, X)$).

Proof. Note that if the condition (C5) holds, then for any bounded subset $\mathcal{Q} \subseteq X$ and for each $\epsilon > 0$, there exists a constant $\delta = (\epsilon)^{\frac{1}{\vartheta}}$ such that for all $x, y \in \mathcal{Q}$ with $\|x - y\| \leq \delta = (\epsilon)^{\frac{1}{\vartheta}}$ and $t \in \mathbb{R}$,

$$\|f(t, x) - f(t, y)\| \leq \mathbb{L}(t)\|x - y\|^{\vartheta} \leq \mathbb{L}(t)[(\epsilon)^{\frac{1}{\vartheta}}]^{\vartheta} < \mathbb{L}(t)\epsilon.$$

Thus the conclusions hold true by Theorem 2.4 (respectively, Corollary 2.5). $\qquad \square$

2.4 *S*-Class Asymptotically Bloch-Type Periodic Functions

In this section, we mainly introduce notions of S-asymptotically Bloch-type periodic function, (weighted) pseudo S-asymptotically Bloch-type periodic function and their fundamental properties (see also Chang and Wei 2021a, 2021b; Chen *et al.*, 2022).

Definition 2.5. A function $f \in BC(\mathbb{R}, X)$ is said to be S-asymptotically ω-periodic if there exists $\omega \in \mathbb{R}$ such that

$$\lim_{|t| \to \infty} \|f(t + \omega) - f(t)\| = 0.$$

The set of such functions will be denoted by $SAP_{\omega}(\mathbb{R}, X)$.

Definition 2.6. A function $f \in BC(\mathbb{R}, X)$ is said to be pseudo S-asymptotically ω-periodic if for a given $\omega \in \mathbb{R}$,

$$\lim_{r \to \infty} \frac{1}{2r} \int_{-r}^{r} \|f(t + \omega) - f(t)\| dt = 0$$

holds for each $t \in \mathbb{R}$. The collection of such functions will be denoted by $PSAP_{\omega}(\mathbb{R}, X)$.

Definition 2.7. A function $f \in BC(\mathbb{R}, X)$ is said to be S-asymptotically Bloch-type periodic (or S-asymptotically (ω, k)-Bloch periodic) if for given $\omega, k \in \mathbb{R}$,

$$\lim_{|t| \to \infty} \left\| f(t + \omega) - e^{ik\omega} f(t) \right\| = 0$$

holds for each $t \in \mathbb{R}$. We denote the space of all such functions by $SABP_{\omega,k}(\mathbb{R}, X)$.

If $k\omega = 2\pi$, Definition 2.7 is reduced to Definition 2.5. Let $k\omega = \pi$ in Definition 2.7, we can also obtain the following notion of S-asymptotically ω-antiperiodic functions.

Definition 2.8. A function $f \in BC(\mathbb{R}, X)$ is said to be S-asymptotically ω-antiperiodic if for a given $\omega \in \mathbb{R}$,

$$\lim_{|t| \to \infty} \left\| f(t + \omega) + f(t) \right\| = 0.$$

The collection of such functions will be denoted by $SAAP_{\omega}(\mathbb{R}, X)$.

Lemma 2.15. *Let* $f_1, f_2, f \in SABP_{\omega,k}(\mathbb{R}, X)$. *Then the following results hold:*

(I) $f_1 + f_2 \in SABP_{\omega,k}(\mathbb{R}, X)$, *and* $cf \in SABP_{\omega,k}(\mathbb{R}, X)$ *for each* $c \in \mathbb{C}$.
(II) *The space* $SABP_{\omega,k}(\mathbb{R}, X)$ *is a Banach space with the sup-norm.*

Proof. (I) From the Definition 2.7, for any $\varepsilon > 0$, there exist constants $\omega, k \in \mathbb{R}$ and $T_\varepsilon > 0$ such that

$$\left\| f(t + \omega) - e^{ik\omega} f(t) \right\| \le \frac{\varepsilon}{|c|},$$

$$\left\| f_i(t + \omega) - e^{ik\omega} f_i(t) \right\| \le \frac{\varepsilon}{2}, \quad i = 1, 2,$$

for each $|t| \ge T_\varepsilon$. Hence,

$$\left\| cf(t + \omega) - ce^{ik\omega} f(t) \right\| \le |c| \left\| f(t + \omega) - e^{ik\omega} f(t) \right\| \le \varepsilon,$$

and

$$\left\| f_1(t + \omega) + f_2(t + \omega) - e^{ik\omega} \left(f_1(t) + f_2(t) \right) \right\|$$
$$\le \left\| f_1(t + \omega) - e^{ik\omega} f_1(t) \right\| + \left\| f_2(t + \omega) - e^{ik\omega} f_2(t) \right\|$$
$$\le \frac{\varepsilon}{2} + \frac{\varepsilon}{2} = \varepsilon.$$

The above arguments imply that $f_1 + f_2, cf \in SABP_{\omega,k}(\mathbb{R}, X)$.

(II) Let $\{f_n\}_n \subseteq SABP(\mathbb{R}, X)$ converge to f as $n \to \infty$. Then for any $\varepsilon > 0$, we can choose suitable constants $N > 0$ and $T_\varepsilon > 0$ such that

$$\left\| f_n(t + \omega) - e^{ik\omega} f_n(t) \right\| \leq \frac{\varepsilon}{3}, \quad \| f_n(t) - f(t) \| \leq \frac{\varepsilon}{3},$$

for $n > N$ and $|t| > T_\varepsilon$. Thus,

$$\begin{aligned}
\big\| f(t + \omega) &- e^{ik\omega} f(t) \big\| \\
&= \big\| f(t + \omega) - f_n(t + \omega) + f_n(t + \omega) - e^{ik\omega} f_n(t) \\
&\quad + e^{ik\omega} f_n(t) - e^{ik\omega} f(t) \big\| \\
&\leq \| f(t + \omega) - f_n(t + \omega) \| + \big\| f_n(t + \omega) - e^{ik\omega} f_n(t) \big\| \\
&\quad + \| f_n(t) - f(t) \| \\
&\leq \frac{\varepsilon}{3} + \frac{\varepsilon}{3} + \frac{\varepsilon}{3} = \varepsilon,
\end{aligned}$$

which implies that the space $SABP_{\omega,k}(\mathbb{R}, X)$ is a closed sub-space of $BC(\mathbb{R}, X)$, so it is a Banach space equipped with the sup-norm. $\qquad\square$

Theorem 2.5. *Let $f \in BC(\mathbb{R} \times X, Y)$ satisfy the following conditions:*

(T1) *For all $(t, x) \in \mathbb{R} \times X$,*

$$f(t + \omega, x) = e^{ik\omega} f\left(t, e^{-ik\omega} x\right).$$

(T2) *There exists a constant $L > 0$ such that for all $x, y \in X$ and $t \in \mathbb{R}$,*

$$\| f(t, x) - f(t, y) \|_Y \leq L \| x - y \|_X.$$

Then for each $\phi \in SABP_{\omega,k}(\mathbb{R}, X)$, $f(\cdot, \phi(\cdot)) \in SABP_{\omega,k}(\mathbb{R}, Y)$.

Proof. Since $\phi \in SABP_{\omega,k}(\mathbb{R}, X)$, for any $\varepsilon > 0$, there exists $T_\varepsilon > 0$ such that for each $|t| \geq T_\varepsilon$, we have $\left\| \phi(t + \omega) - e^{ik\omega} \phi(t) \right\|_X \leq \frac{\varepsilon}{L}$. On the other hand,

$$\begin{aligned}
\big\| f(t + \omega, \phi(t + \omega)) &- e^{ik\omega} f(t, \phi(t)) \big\|_Y \\
&= \big\| e^{ik\omega} f\left(t, e^{-ik\omega} \phi(t + \omega)\right) - e^{ik\omega} f(t, \phi(t)) \big\|_Y \quad \text{(owing to (T1))} \\
&\leq L \big\| e^{-ik\omega} \phi(t + \omega) - \phi(t) \big\|_X, \quad \text{(owing to (T2))}.
\end{aligned}$$

Thus,

$$\big\| f(t + \omega, \phi(t + \omega)) - e^{ik\omega} f(t, \phi(t)) \big\|_Y \leq \varepsilon$$

for any $\varepsilon > 0$ and each $|t| > T_\varepsilon$. $\qquad\square$

Theorem 2.6. *Let* $f \in BC(\mathbb{R} \times X, Y)$ *satisfy (T1) and the following condition:*

(T3) *For any* $\varepsilon > 0$ *and any bounded subset* $\mathcal{Q} \subseteq X$, *there exist constants* $T_{\varepsilon,\mathcal{Q}} \geq 0$ *and* $\delta_{\varepsilon,\mathcal{Q}} > 0$ *such that*

$$\|f(t,x) - f(t,y)\|_Y \leq \varepsilon$$

for all $x, y \in \mathcal{Q}$ *with* $\|x - y\| \leq \delta_{\varepsilon,\mathcal{Q}}$ *and* $|t| \geq T_{\varepsilon,\mathcal{Q}}$.

Then for each $\phi \in SABP_{\omega,k}(\mathbb{R}, X)$, $f(\cdot, \phi(\cdot)) \in SABP_{\omega,k}(\mathbb{R}, Y)$.

Proof. Taking into account that

$$\left\|f(t + \omega, \phi(t + \omega)) - e^{ik\omega} f(t, \phi(t))\right\|_Y$$
$$= \left\|f\left(t, e^{-ik\omega}\phi(t + \omega)\right) - f(t, \phi(t))\right\|_Y.$$

Since $\phi \in SABP_{\omega,k}(\mathbb{R}, X)$, then for any $\varepsilon > 0$, there exists $T_\varepsilon > 0$ such that for each $|t| \geq T_\varepsilon$,

$$\left\|e^{-ik\omega}\phi(t + \omega) - \phi(t)\right\|_X \leq \varepsilon.$$

Based upon the condition (T3), for any $\varepsilon > 0$, there exists $\delta_{\varepsilon,\mathcal{Q}} := \varepsilon$ and $T_{\varepsilon,\mathcal{Q}} := T_\varepsilon$ such that

$$\left\|f\left(t, e^{-ik\omega}\phi(t + \omega)\right) - f(t, \phi(t))\right\|_Y \leq \varepsilon$$

for each $|t| > T_{\varepsilon,\mathcal{Q}}$, i.e.,

$$\left\|f(t + \omega, \phi(t + \omega)) - e^{ik\omega} f(t, \phi(t))\right\|_Y \leq \varepsilon$$

for each $|t| > T_{\varepsilon,\mathcal{Q}}$. $\square$

Remark 2.4. The condition (T1) in Theorems 2.5–2.6 can be weakened by the following condition:

(T1′) *For all* $(t, x) \in \mathbb{R} \times X$, $\displaystyle\lim_{|t|\to\infty} \left\|f(t + \omega, x) - e^{ik\omega} f\left(t, e^{-ik\omega}x\right)\right\|_Y = 0$
    uniformly on any bounded set of X.

In fact, if (T1′) and (T2) hold, then for any $\varepsilon > 0$, there exists T_ε^1 such that for each $|t| \geq T_\varepsilon^1$, we have

$$\left\|f(t + \omega, x) - e^{ik\omega} f\left(t, e^{-ik\omega}x\right)\right\|_Y \leq \frac{\varepsilon}{2}.$$

Meanwhile, for each $\phi \in SABP_{\omega,k}(\mathbb{R}, X)$, there exists $T_\varepsilon^2 > 0$ such that for each $|t| \geq T_\varepsilon^2$, we have

$$\left\| \phi(t + \omega) - e^{ik\omega} \phi(t) \right\|_X \leq \frac{\varepsilon}{2L}.$$

Thus, there exists $T_\varepsilon = \max\{T_\varepsilon^1, T_\varepsilon^2\}$ such that for each $|t| \geq T_\varepsilon$, we have

$$\left\| f(t + \omega, \phi(t + \omega)) - e^{ik\omega} f(t, \phi(t)) \right\|_Y$$
$$= \left\| f(t + \omega, \phi(t + \omega)) - e^{ik\omega} f\left(t, e^{-ik\omega}\phi(t+\omega)\right) \right.$$
$$\left. + e^{ik\omega} f\left(t, e^{-ik\omega}\phi(t+\omega)\right) - e^{ik\omega} f(t, \phi(t)) \right\|$$
$$\leq \left\| f(t + \omega, \phi(t + \omega)) - e^{ik\omega} f\left(t, e^{-ik\omega}\phi(t+\omega)\right) \right\|_Y$$
$$+ \left\| f\left(t, e^{-ik\omega}\phi(t+\omega)\right) - f(t, \phi(t)) \right\|_Y$$
$$\leq \frac{\varepsilon}{2} + L \left\| e^{-ik\omega}\phi(t+\omega) - \phi(t) \right\|_X \leq \frac{\varepsilon}{2} + \frac{\varepsilon}{2} = \varepsilon.$$

Thus, the conclusion of Theorem 2.5 still holds true under conditions (T1′) and (T2). The case for (T1′) and (T3) holding can be similarly proved. Particularly, let $k\omega = \pi$, we can obtain some useful composition theorems for S-asymptotically ω-anti-periodic functions under conditions (T1)((T1′))-(T2) or (T1)((T1′)) and (T3) by virtue of Theorems 2.5–2.6. On the other hand, if $k\omega = 2\pi$ and conditions (T1′), (T3) hold, then the conclusion of Theorem 2.6 is just a slightly modification of Henríquez *et al.* (2008a, Lemma 4.1) for S-asymptotically ω-periodic functions.

Lemma 2.16. *Let the condition (T2) (or (T3)) holds. Assume further that $f \in BC(\mathbb{R} \times X, Y)$ is S-asymptotically ω-antiperiodic for all $t \in \mathbb{R}$ uniformly on any bounded set of X, and $\phi \in SAP_\omega(\mathbb{R}, X)$, then the function $f(\cdot, \phi(\cdot)) \in SAAP_\omega(\mathbb{R}, Y)$.*

Proof. Let (T2) hold, then

$$\left\| f(t + \omega, \phi(t + \omega)) + f(t, \phi(t)) \right\|_Y$$
$$= \left\| f(t + \omega, \phi(t + \omega)) + f\left(t, \phi(t + \omega)\right) \right.$$
$$\left. - f\left(t, \phi(t + \omega)\right) + f(t, \phi(t)) \right\|_Y$$
$$\leq \left\| f(t + \omega, \phi(t + \omega)) + f\left(t, \phi(t + \omega)\right) \right\|_Y$$
$$+ \left\| f\left(t, \phi(t + \omega)\right) - f(t, \phi(t)) \right\|_Y$$
$$\leq \left\| f(t + \omega, \phi(t + \omega)) + f\left(t, \phi(t + \omega)\right) \right\|_Y$$
$$+ L \left\| \phi(t + \omega) - \phi(t) \right\|_X. \tag{2.2}$$

Taking into account that f is S-asymptotically ω-antiperiodic for all $t \in \mathbb{R}$ uniformly on any bounded set of X and $\phi \in SAP_\omega(\mathbb{R}, X)$, then for any $\varepsilon > 0$, there exists a constant $T_\varepsilon > 0$ such that for each $|t| \geq T_\varepsilon$

$$\|f(t + \omega, \phi(t + \omega)) + f(t, \phi(t + \omega))\|_Y \leq \frac{\varepsilon}{2}, \ \|\phi(t + \omega) - \phi(t)\|_X \leq \frac{\varepsilon}{2L}.$$

Thus, it follows from (2.2) that

$$\|f(t + \omega, \phi(t + \omega)) + f(t, \phi(t))\|_Y \leq \varepsilon$$

for each $|t| \geq T_\varepsilon$, i.e., $f(\cdot, \phi(\cdot))$ is S-asymptotically ω-antiperiodic. The case for (T3) holding can be similarly conducted by noting that the range of ϕ is bounded. $\qquad\square$

Theorem 2.7. *Let* $\{S(t)\}_{t \geq 0} \subseteq \mathcal{B}(X)$ *be a uniformly integrable and strongly continuous family. If* $f \in SABP_{\omega,k}(\mathbb{R}, X)$, *then*

$$u(t) := \int_{-\infty}^{t} S(t - s) f(s) ds \in SABP_{\omega,k}(\mathbb{R}, X).$$

Proof. Taking into account that the operator family $\{S(t)\}_{t \geq 0} \subseteq \mathcal{B}(X)$ is assumed to be uniformly integrable, we have

$$\int_0^\infty \|S(t)\| dt < \infty.$$

Since $f \in SABP_{\omega,k}(\mathbb{R}, X)$, for any $\varepsilon > 0$, there exists a positive constant T_ε such that $\|f(t + \omega) - e^{ik\omega} f(t)\| \leq \varepsilon$ for each $t > T_\varepsilon$. On the other hand,

$$\left\| u(t + \omega) - e^{i\omega k} u(t) \right\|$$

$$= \left\| \int_{-\infty}^{t+\omega} S(t + \omega - s) f(s) ds - \int_{-\infty}^{t} S(t - s) e^{i\omega k} f(s) ds \right\|$$

$$= \left\| \int_{-\infty}^{t} S(t - s) \left[f(s + \omega) - e^{i\omega k} f(s) \right] ds \right\|$$

$$\leq \int_{-\infty}^{t} \left\| S(t - s) \left[f(s + \omega) - e^{i\omega k} f(s) \right] \right\| ds$$

$$\leq \int_{-\infty}^{T_\varepsilon} \left\| S(t - s) \left[f(s + \omega) - e^{i\omega k} f(s) \right] \right\| ds$$

$$+ \int_{T_\varepsilon}^{t} \left\| S(t - s) \left[f(s + \omega) - e^{i\omega k} f(s) \right] \right\| ds$$

$$\leq 2\|f\|_\infty \int_{-\infty}^{T_\varepsilon} \|\mathcal{S}(t-s)\|ds + \varepsilon \int_{T_\varepsilon}^{t} \|\mathcal{S}(t-s)\|ds$$

$$\leq 2\|f\|_\infty \int_{t-T_\varepsilon}^{\infty} \|\mathcal{S}(s)\|ds + \varepsilon \int_{0}^{\infty} \|\mathcal{S}(s)\|ds,$$

which implies that $\|u(t+\omega) - e^{i\omega k}u(t)\| \to 0$ as $t \to \infty$. $\qquad\square$

Remark 2.5. If $k\omega = \pi$, then we can obtain a useful convolution property for S-asymptotically ω-antiperiodic functions by Theorem 2.7.

Definition 2.9. A function $f \in BC(\mathbb{R}, X)$ is said to be pseudo S-asymptotically Bloch-type periodic (or pseudo S-asymptotically (ω, k)-Bloch periodic) if for given $\omega, k \in \mathbb{R}$,

$$\lim_{r\to\infty} \frac{1}{2r} \int_{-r}^{r} \left\| f(t+\omega) - e^{ik\omega}f(t) \right\| dt = 0$$

holds for each $t \in \mathbb{R}$. We denote the space of all such functions by $PSABP_{\omega,k}(\mathbb{R}, X)$.

If $k\omega = 2\pi$, Definition 2.9 is reduced to Definition 2.6. Let $k\omega = \pi$ in Definition 2.9, we can also obtain the following definition of pseudo S-asymptotically ω-antiperiodic functions.

Definition 2.10. A function $f \in BC(\mathbb{R}, X)$ is called to be a pseudo S-asymptotically ω-antiperiodic function if for a given $\omega \in \mathbb{R}$,

$$\lim_{r\to\infty} \frac{1}{2r} \int_{-r}^{r} \|f(t+\omega) + f(t)\|dt = 0$$

holds for each $t \in \mathbb{R}$. The collection of such functions will be denoted by $PSAAP_\omega(\mathbb{R}, X)$.

Lemma 2.17. *Let* $f_1, f_2, f \in PSABP_{\omega,k}(\mathbb{R}, X)$. *Then the following results hold:*

(I) $f_1 + f_2 \in PSABP_{\omega,k}(\mathbb{R}, X)$, *and* $cf \in PSABP_{\omega,k}(\mathbb{R}, X)$ *for each* $c \in \mathbb{C}$.

(II) $f_a \in PSABP_{\omega,k}(\mathbb{R}, X)$ *for each* $a \in \mathbb{R}$.

(III) *The space* $PSABP_{\omega,k}(\mathbb{R}, X)$ *is a Banach space with the sup-norm.*

Proof. (I) By Definition 2.9, we have

$$\frac{1}{2r} \int_{-r}^{r} \left\| f(t+\omega) - e^{ik\omega} f(t) \right\| dt \to 0,$$

$$\frac{1}{2r} \int_{-r}^{r} \left\| f_i(t+\omega) - e^{ik\omega} f_i(t) \right\| dt \to 0, \ i = 1, 2,$$

as $r \to \infty$. Hence,

$$\frac{1}{2r} \int_{-r}^{r} \left\| cf(t+\omega) - ce^{ik\omega} f(t) \right\| dt$$

$$\leq |c| \frac{1}{2r} \int_{-r}^{r} \left\| f(t+\omega) - e^{ik\omega} f(t) \right\| dt \to 0,$$

and

$$\frac{1}{2r} \int_{-r}^{r} \left\| f_1(t+\omega) + f_2(t+\omega) - e^{ik\omega} \left(f_1(t) + f_2(t) \right) \right\| dt$$

$$\leq \frac{1}{2r} \int_{-r}^{r} \left\| f_1(t+\omega) - e^{ik\omega} f_1(t) \right\| dt$$

$$+ \frac{1}{2r} \int_{-r}^{r} \left\| f_2(t+\omega) - e^{ik\omega} f_2(t) \right\| dt$$

$$\to 0$$

as $r \to \infty$. Thus, we have the claim.

(II) Let $f \in PSABP_{\omega,k}(\mathbb{R}, X)$, then by the Definition 2.9, we have

$$\frac{1}{2r} \int_{-r}^{r} \left\| f(t+\omega) - e^{ik\omega} f(t) \right\| dt \to 0,$$

as $r \to \infty$. Thus, for each $a \in \mathbb{R}$

$$\frac{1}{2r} \int_{-r}^{r} \left\| f(t+a+\omega) - e^{ik\omega} f(t+a) \right\| dt$$

$$= \frac{1}{2r} \int_{-r+a}^{r+a} \left\| f(t+\omega) - e^{ik\omega} f(t) \right\| dt$$

$$\leq \frac{1}{2r} \int_{-r-|a|}^{r+|a|} \left\| f(t+\omega) - e^{ik\omega} f(t) \right\| dt$$

$$= \frac{2(r+|a|)}{2r} \frac{1}{2(r+|a|)} \int_{-r-|a|}^{r+|a|} \left\| f(t+\omega) - e^{ik\omega} f(t) \right\| dt \to 0$$

as $r \to \infty$, i.e., the translated $f_a \in PSABP_{\omega,k}(\mathbb{R}, X)$.

(III) Let $\{f_n\}_n \subseteq PSABP_{\omega,k}(\mathbb{R}, X)$ converge to f as $n \to \infty$. Then for any $\varepsilon > 0$, we can choose suitable constants $N > 0$ and $r_\varepsilon > 0$ such that

$$\frac{1}{2r} \int_{-r}^{r} \left\| f_n(t+\omega) - e^{ik\omega} f_n(t) \right\| dt \leq \frac{\varepsilon}{3}, \quad \|f_n - f\|_\infty \leq \frac{\varepsilon}{3},$$

for $n > N$ and $r > r_\varepsilon$. Thus,

$$\frac{1}{2r} \int_{-r}^{r} \left\| f(t+\omega) - e^{ik\omega} f(t) \right\| dt$$

$$= \frac{1}{2r} \int_{-r}^{r} \left\| f(t+\omega) - f_n(t+\omega) + f_n(t+\omega) - e^{ik\omega} f_n(t) \right.$$

$$\left. + e^{ik\omega} f_n(t) - e^{ik\omega} f(t) \right\| dt$$

$$\leq \|f - f_n\|_\infty + \frac{1}{2r} \int_{-r}^{r} \left\| f_n(t+\omega) - e^{ik\omega} f_n(t) \right\| dt$$

$$+ \|f_n - f\|_\infty$$

$$\leq \frac{\varepsilon}{3} + \frac{\varepsilon}{3} + \frac{\varepsilon}{3} = \varepsilon,$$

which implies that the space $PSABP_{\omega,k}(\mathbb{R}, X)$ is a closed sub-space of $BC(\mathbb{R}, X)$, so it is a Banach space equipped with the sup-norm. $\quad\square$

Theorem 2.8. *Let $f \in BC(\mathbb{R} \times X, Y)$ satisfy (T1)–(T2) in Theorem 2.5. Then for each $\phi \in PSABP_{\omega,k}(\mathbb{R}, X)$, we have*

$$f(\cdot, \phi(\cdot)) \in PSABP_{\omega,k}(\mathbb{R}, Y).$$

Proof. In view of $\phi \in PSABP_{\omega,k}(\mathbb{R}, X)$, for each $t \in \mathbb{R}$, we have

$$\frac{1}{2r} \int_{-r}^{r} \left\| \phi(t+\omega) - e^{ik\omega} \phi(t) \right\| dt \to 0$$

as $r \to \infty$. On the other hand,

$$\frac{1}{2r} \int_{-r}^{r} \left\| f(t+\omega, \phi(t+\omega)) - e^{ik\omega} f(t, \phi(t)) \right\| dt$$

$$= \frac{1}{2r} \int_{-r}^{r} \left\| e^{ik\omega} f\left(t, e^{-ik\omega} \phi(t+\omega)\right) - e^{ik\omega} f(t, \phi(t)) \right\| dt$$

$$\leq L \frac{1}{2r} \int_{-r}^{r} \left\| e^{-ik\omega} \phi(t+\omega) - \phi(t) \right\| dt.$$

Thus,

$$\frac{1}{2r} \int_{-r}^{r} \left\| f(t+\omega, \phi(t+\omega)) - e^{ik\omega} f(t, \phi(t)) \right\| dt \to 0$$

as $r \to \infty$, i.e., $f(\cdot, \phi(\cdot)) \in PSABP_{\omega,k}(\mathbb{R}, Y)$ by Definition 2.9. $\qquad\square$

Lemma 2.18. *If $f \in BC(\mathbb{R}, X)$, then the following assertions are equivalent:*

(a) $\lim_{r\to\infty} \frac{1}{2r} \int_{-r}^{r} \left\| f(t+\omega) - e^{ik\omega} f(t) \right\| dt = 0.$
(b) *For each $\epsilon > 0$,*

$$\lim_{r\to\infty} \frac{1}{2r} \int_{M_{r,\epsilon}(f)} dt = 0,$$

where $M_{r,\epsilon}(f) = \{t \in [-r, r] : \left\| f(t+\omega) - e^{ik\omega} f(t) \right\| \geq \epsilon\}$.

Proof. For each given $\epsilon > 0$, since

$$\frac{1}{2r} \int_{-r}^{r} \left\| f(t+\omega) - e^{ik\omega} f(t) \right\| dt$$

$$= \frac{1}{2r} \int_{[-r,r] \backslash M_{r,\epsilon}(f)} \left\| f(t+\omega) - e^{ik\omega} f(t) \right\| dt$$

$$+ \frac{1}{2r} \int_{M_{r,\epsilon}(f)} \left\| f(t+\omega) - e^{ik\omega} f(t) \right\| dt$$

$$\geq \frac{1}{2r} \int_{M_{r,\epsilon}(f)} \left\| f(t+\omega) - e^{ik\omega} f(t) \right\| dt$$

$$\geq \frac{\epsilon}{2r} \int_{M_{r,\epsilon}(f)} dt \geq 0,$$

we can deduce the assertion (b) if the assertion (a) is true. On the other hand,

$$\frac{1}{2r} \int_{-r}^{r} \left\| f(t+\omega) - e^{ik\omega} f(t) \right\| dt$$

$$\leq \left(1 - \frac{\int_{M_{r,\epsilon}(f)} dt}{2r} \right) \epsilon + \frac{1}{2r} 2\|f\|_{\infty} \int_{M_{r,\epsilon}(f)} dt,$$

we can obtain the assertion (a) by the truth of the assertion (b). □

Theorem 2.9. *Let* $f \in BC(\mathbb{R} \times X, Y)$ *satisfy (T1) and (T3) in Theorem 2.6. Then for each* $\phi \in PSABP_{\omega,k}(\mathbb{R}, X)$, $f(\cdot, \phi(\cdot)) \in PSABP_{\omega,k}(\mathbb{R}, Y)$.

Proof. The condition (T1) gives that

$$\left\| f(t+\omega, \phi(t+\omega)) - e^{ik\omega} f(t, \phi(t)) \right\|$$

$$= \left\| f\left(t, e^{-ik\omega} \phi(t+\omega) \right) - f(t, \phi(t)) \right\|.$$

Since $\phi \in PSABP_{\omega,k}(\mathbb{R}, X)$ and by Lemma 2.18, for any $\varepsilon > 0$, there exists $r_\varepsilon > 0$ such that for each $r \geq r_\varepsilon$,

$$\frac{1}{2r} \int_{M_{r,\epsilon}(\phi)} dt \leq \frac{\varepsilon}{4\|f\|}.$$

Throught the condition (T3), for any $\varepsilon > 0$, there exists $\delta_{\varepsilon,\mathcal{Q}} := \varepsilon$ and $T_{\varepsilon,\mathcal{Q}} = r_\varepsilon$ such that $\left\| f\left(t, e^{-ik\omega} \phi(t+\omega) \right) - f(t, \phi(t)) \right\| \leq \varepsilon$ whenever $\left\| \phi(t+\omega) - e^{ik\omega} \phi(t) \right\| \leq \varepsilon$ and $|t| \geq r_\varepsilon$. Hence,

$$\frac{1}{2r} \int_{-r}^{r} \left\| f\left(t, e^{-ik\omega} \phi(t+\omega) \right) - f(t, \phi(t)) \right\| dt$$

$$= \frac{1}{2r} \int_{[-r,r] \backslash M_{r,\epsilon}(\phi)} \left\| f\left(t, e^{-ik\omega} \phi(t+\omega) \right) - f(t, \phi(t)) \right\| dt$$

$$+ \frac{1}{2r} \int_{M_{r,\epsilon}(\phi)} \left\| f\left(t, e^{-ik\omega} \phi(t+\omega) \right) - f(t, \phi(t)) \right\| dt$$

$$\leq \left(1 - \frac{\int_{M_{r,\epsilon}(\phi)} dt}{2r} \right) \varepsilon + \frac{1}{2r} 2\|f\| \int_{M_{r,\epsilon}(\phi)} dt \leq \frac{3\varepsilon}{2}$$

for each $r \geq r_\varepsilon$, i.e.,

$$\frac{1}{2r} \int_{-r}^{r} \left\| f(t+\omega, \phi(t+\omega)) - e^{ik\omega} f(t, \phi(t)) \right\| dt \to 0$$

as $r \to \infty$, i.e., $f(\cdot, \phi(\cdot)) \in PSABP_{\omega,k}(\mathbb{R}, Y)$ by Definition 2.9. $\qquad \square$

Remark 2.6. The condition (T1) in Theorems 2.8 and 2.9 can be replaced by the following condition:

(T1*) For all $(t, x) \in \mathbb{R} \times X$,

$$\lim_{r \to \infty} \frac{1}{2r} \int_{-r}^{r} \left\| f(t+\omega, x) - e^{ik\omega} f\left(t, e^{-ik\omega} x\right) \right\| dt = 0$$

uniformly on any bounded set of X.

In fact, if (T1*) and (T2) hold, then for each $t \in \mathbb{R}$ and $\phi \in PSABP_{\omega,k}(\mathbb{R}, X)$, we have

$$\frac{1}{2r} \int_{-r}^{r} \left\| f(t+\omega, \phi(t+\omega)) - e^{ik\omega} f(t, \phi(t)) \right\| dt$$

$$= \frac{1}{2r} \int_{-r}^{r} \left\| f(t+\omega, \phi(t+\omega)) - e^{ik\omega} f(t, e^{-ik\omega} \phi(t+\omega)) \right.$$

$$\left. + e^{ik\omega} f(t, e^{-ik\omega} \phi(t+\omega)) - e^{ik\omega} f(t, \phi(t)) \right\| dt$$

$$\leq \frac{1}{2r} \int_{-r}^{r} \left\| f(t+\omega, \phi(t+\omega)) - e^{ik\omega} f\left(t, e^{-ik\omega} \phi(t+\omega)\right) \right\| dt$$

$$+ \frac{1}{2r} \int_{-r}^{r} \left\| f\left(t, e^{-ik\omega} \phi(t+\omega)\right) - f(t, \phi(t)) \right\| dt$$

$$\leq \frac{1}{2r} \int_{-r}^{r} \left\| f(t+\omega, \phi(t+\omega)) - e^{ik\omega} f\left(t, e^{-ik\omega} \phi(t+\omega)\right) \right\| dt$$

$$+ L \frac{1}{2r} \int_{-r}^{r} \left\| e^{-ik\omega} \phi(t+\omega) - \phi(t) \right\| dt$$

$$\to 0$$

as $r \to \infty$. Thus, the assertion of Theorem 2.8 is still true under conditions (T1*) and (T2). Similarly, the conclusion of Theorem 2.9 holds true if conditions (T1*) and (T3) are satisfied.

Lemma 2.19. *Let the condition* (T2) (*or* (T3)) *hold. Assume further that* $f \in BC(\mathbb{R} \times X, Y)$ *is pseudo S-asymptotically ω-antiperiodic for all $t \in \mathbb{R}$ uniformly on any bounded set of X. Then for each $\phi \in PSAP_\omega(\mathbb{R}, X)$, the function $f(\cdot, \phi(\cdot)) \in PSAAP_\omega(\mathbb{R}, Y)$.*

Proof. Since the condition (T2) holds, we have

$$\|f(t+\omega, \phi(t+\omega)) + f(t, \phi(t))\|$$

$$= \|f(t+\omega, \phi(t+\omega)) + f(t, \phi(t+\omega))$$

$$- f(t, \phi(t+\omega)) + f(t, \phi(t))\|$$

$$\leq \|f(t+\omega, \phi(t+\omega)) + f(t, \phi(t+\omega))\|$$

$$+ \|f(t, \phi(t+\omega)) - f(t, \phi(t))\|$$

$$\leq \|f(t+\omega, \phi(t+\omega)) + f(t, \phi(t+\omega))\|$$

$$+ L \|\phi(t+\omega) - \phi(t)\|. \tag{2.3}$$

Because f is pseudo S-asymptotically ω-antiperiodic for all $t \in \mathbb{R}$ uniformly on any bounded set of X and $\phi \in PSAP_\omega(\mathbb{R}, X)$, we have

$$\frac{1}{2r} \int_{-r}^{r} \|f(t+\omega, \phi(t+\omega)) + f(t, \phi(t+\omega))\| dt \to 0,$$

$$\frac{1}{2r} \int_{-r}^{r} \|\phi(t+\omega) - \phi(t)\| dt \to 0$$

as $r \to \infty$. Thus, it follows from the relation (2.3) that

$$\frac{1}{2r} \int_{-r}^{r} \|f(t+\omega, \phi(t+\omega)) + f(t, \phi(t))\| dt \to 0$$

as $r \to \infty$, i.e., $f(\cdot, \phi(\cdot))$ is pseudo S-asymptotically ω-antiperiodic. The case for (T3) holding can be similarly conducted by noting that the range of ϕ is bounded. $\qquad \square$

Theorem 2.10. *Let $\{\mathcal{S}(t)\}_{t\geq 0} \subseteq \mathcal{B}(X)$ be a uniformly integrable and strongly continuous family. If $f \in PSABP_{\omega,k}(\mathbb{R}, X)$, then*

$$u(t) := \int_{-\infty}^{t} \mathcal{S}(t-s)f(s)ds \in PSABP_{\omega,k}(\mathbb{R}, X).$$

Proof. Since the operator family $\{\mathcal{S}(t)\}_{t\geq0} \subseteq \mathcal{B}(X)$ is assumed to be uniformly integrable, we have

$$\int_0^\infty \|\mathcal{S}(t)\| dt < \infty.$$

Owing to $f \in PSABP_{\omega,k}(\mathbb{R}, X)$, for any $t \in \mathbb{R}$, we have

$$\frac{1}{2r} \int_{-r}^r \|f(t+\omega) - e^{ik\omega} f(t)\| dt \to 0 \tag{2.4}$$

as $r \to \infty$. On the other hand, by the Fubini theorem, we have

$$\frac{1}{2r} \int_{-r}^r \|u(t+\omega) - e^{i\omega k} u(t)\| dt$$

$$= \frac{1}{2r} \int_{-r}^r \left\| \int_{-\infty}^{t+\omega} \mathcal{S}(t+\omega-s) f(s) ds \right.$$

$$\left. - \int_{-\infty}^t \mathcal{S}(t-s) e^{i\omega k} f(s) ds \right\| dt$$

$$= \frac{1}{2r} \int_{-r}^r \left\| \int_{-\infty}^t \mathcal{S}(t-s) \left[f(s+\omega) - e^{i\omega k} f(s) \right] ds \right\| dt$$

$$\leq \frac{1}{2r} \int_{-r}^r \left[\int_{-\infty}^t \|\mathcal{S}(t-s) \left[f(s+\omega) - e^{i\omega k} f(s) \right]\| ds \right] dt$$

$$\leq \frac{1}{2r} \int_{-r}^r \left[\int_0^\infty \|\mathcal{S}(s) \left[f(t-s+\omega) - e^{i\omega k} f(t-s) \right]\| ds \right] dt$$

$$= \int_0^\infty \|\mathcal{S}(s)\| \left[\frac{1}{2r} \int_{-r}^r \|f(t-s+\omega) - e^{i\omega k} f(t-s)\| dt \right] ds.$$

Through the relation (2.4), Lemma 2.17 (II) and the Lebesgue dominated convergence theorem, we have

$$\frac{1}{2r} \int_{-r}^r \|u(t+\omega) - e^{i\omega k} u(t)\| dt \to 0$$

as $r \to \infty$, i.e., $u \in PSABP_{\omega,k}(\mathbb{R}, X)$. $\qquad\square$

Remark 2.7. If $k\omega = \pi$, we can obtain some basic properties for pseudo S-asymptotically ω-antiperiodic functions by Lemmas 2.17–2.18, Theorems 2.8–2.10 and Remark 2.6.

Definition 2.11. Let $\rho \in \mathcal{U}_\infty$, a function $f \in BC(\mathbb{R}, X)$ is said to be weighted pseudo S-asymptotically Bloch-type periodic (or weighted pseudo S-asymptotically (ω, k)-Bloch periodic) if for given $\omega, k \in \mathbb{R}$,

$$\lim_{r\to\infty} \frac{1}{\mu(r, \rho)} \int_{-r}^{r} \left\| f(t+\omega) - e^{ik\omega} f(t) \right\| \rho(t) dt = 0$$

holds for each $t \in \mathbb{R}$. We denote the space of all such functions by $WPSABP_{\omega,k}(\mathbb{R}, X, \rho)$.

Let $k\omega := \pi$ in Definition 2.11, we can also have the following notion of weighted pseudo S-asymptotically ω-antiperiodic functions.

Definition 2.12. Let $\rho \in \mathcal{U}_\infty$, a function $f \in BC(\mathbb{R}, X)$ is called to be a weighted pseudo S-asymptotically ω-antiperiodic function if for a given $\omega \in \mathbb{R}$,

$$\lim_{r\to\infty} \frac{1}{\mu(r, \rho)} \int_{-r}^{r} \| f(t+\omega) + f(t) \| \rho(t) dt = 0$$

holds for each $t \in \mathbb{R}$. The collection of such functions will be denoted by $WPSAAP_\omega(\mathbb{R}, X, \rho)$.

Lemma 2.20. *Let $\rho \in \mathcal{U}_\infty$ and $f_1, f_2, f \in WPSABP_{\omega,k}(\mathbb{R}, X, \rho)$, then the following results hold:*

(i) $f_1 + f_2 \in WPSABP_{\omega,k}(\mathbb{R}, X, \rho)$, *and* $cf \in WPSABP_{\omega,k}(\mathbb{R}, X, \rho)$ *for each* $c \in \mathbb{C}$.
(ii) *If ρ satisfies H_ρ, then* $f_a \in WPSABP_{\omega,k}(\mathbb{R}, X, \rho)$ *for each* $a \in \mathbb{R}$.
(iii) *The space $WPSABP_{\omega,k}(\mathbb{R}, X, \rho)$ is a Banach space with the sup-norm.*

Proof. The proof of (i) and (iii) can be conducted similarly as (I) and (III) in Lemma 2.17 by Definition 2.11. For the assertion (ii), the assumption H_ρ implies that there exists a constant $\vartheta > 0$ such that for a.e. $t \in \mathbb{R}$

$$\frac{\rho(t-a)}{\rho(t)} \leq \vartheta, \quad \frac{\mu(r+|a|, \rho)}{\mu(r, \rho)} \leq \vartheta.$$

Thus, for a given $a \in \mathbb{R}$

$$\frac{1}{\mu(r,\rho)} \int_{-r}^{r} \left\| f(t+a+\omega) - e^{ik\omega} f(t+a) \right\| \rho(t) dt$$

$$= \frac{1}{\mu(r,\rho)} \int_{-r+a}^{r+a} \left\| f(t+\omega) - e^{ik\omega} f(t) \right\| \rho(t-a) dt$$

$$\leq \frac{1}{\mu(r,\rho)} \int_{-r-|a|}^{r+|a|} \left\| f(t+\omega) - e^{ik\omega} f(t) \right\| \rho(t-a) dt$$

$$= \frac{\mu(r+|a|,\rho)}{\mu(r,\rho)} \frac{1}{\mu(r+|a|,\rho)}$$

$$\times \int_{-r-|a|}^{r+|a|} \left\| f(t+\omega) - e^{ik\omega} f(t) \right\| \rho(t) \frac{\rho(t-a)}{\rho(t)} dt$$

$$\leq \vartheta^2 \frac{1}{\mu(r+|a|,\rho)} \int_{-r-|a|}^{r+|a|} \left\| f(t+\omega) - e^{ik\omega} f(t) \right\| \rho(t) dt \to 0$$

as $r \to \infty$ due to $f \in WPSABP_{\omega,k}(\mathbb{R}, X, \rho)$, i.e., the translation $f_a \in WPSABP_{\omega,k}(\mathbb{R}, X, \rho)$. $\qquad \square$

Lemma 2.21. *Let $\rho \in \mathcal{U}_\infty$ and $f \in BC(\mathbb{R}, X)$, then the following assertions are equivalent:*

(c) $\lim_{r \to \infty} \frac{1}{\mu(r,\rho)} \int_{-r}^{r} \| f(t+\omega) - e^{ik\omega} f(t) \| \rho(t) dt = 0.$
(d) *For each $\epsilon > 0$,*

$$\lim_{r \to \infty} \frac{1}{\mu(r,\rho)} \int_{M_{r,\epsilon}(f)} \rho(t) dt = 0,$$

where $M_{r,\epsilon}(f)$ is given in Lemma 2.18.

Proof. The proof can be completed similarly to Lemma 2.18 by Definition 2.11. $\qquad \square$

By Definition 2.11 and Lemma 2.20(ii), we can obtain the following results analogously to proofs of Theorems 2.8–2.9.

Theorem 2.11. *Let $\rho \in \mathcal{U}_\infty$ and $f \in BC(\mathbb{R} \times X, Y)$ satisfy conditions (T1)–(T2). Then for each $\phi \in WPSABP_{\omega,k}(\mathbb{R}, X, \rho)$, $f(\cdot, \phi(\cdot)) \in WPSABP_{\omega,k}(\mathbb{R}, Y, \rho)$.*

Theorem 2.12. *Let $\rho \in \mathcal{U}_\infty$ and $f \in BC(\mathbb{R} \times X, Y)$ satisfy conditions (T1) and (T3). Then for each $\phi \in WPSABP_{\omega,k}(\mathbb{R}, X, \rho)$, $f(\cdot, \phi(\cdot)) \in WPSABP_{\omega,k}(\mathbb{R}, Y, \rho)$.*

Remark 2.8. Similarly, for weighted pseudo S-asymptotically Bloch-type periodic functions, the condition (T1) in Theorems 2.11–2.12 can be replaced by the following condition:

(T1″) For all $(t, x) \in \mathbb{R} \times X$,

$$\lim_{r \to \infty} \frac{1}{\mu(r, \rho)} \int_{-r}^{r} \left\| f(t + \omega, x) - e^{ik\omega} f(t, e^{-ik\omega} x) \right\| \rho(t) dt = 0$$

uniformly on any bounded set of X.

By Definition 2.11 and Lemma 2.21, we can have the following result similarly to the proof of Theorem 2.7.

Theorem 2.13. *Let $\{\mathcal{S}(t)\}_{t \geq 0} \subseteq \mathcal{B}(X)$ be a uniformly integrable and strongly continuous family. If $\rho \in \mathcal{U}_{\text{inv}}$, $f \in WPSABP_{\omega,k}(\mathbb{R}, X, \rho)$, then*

$$u(t) := \int_{-\infty}^{t} \mathcal{S}(t - s) f(s) ds \in WPSABP_{\omega,k}(\mathbb{R}, X, \rho).$$

Remark 2.9. Let $k\omega = \pi$, we can deduce some properties on weighted pseudo S-asymptotically ω-antiperiodic functions by Lemmas 2.20–2.21 and Theorems 2.11–2.13.

Similar to Lemma 2.19, we also have the following result.

Lemma 2.22. *Let $\rho \in \mathcal{U}_\infty$ and the condition (T2) (or(T3)) hold. Suppose further that $f \in BC(\mathbb{R} \times X, X)$ is weighted pseudo S-asymptotically ω-antiperiodic for all $t \in \mathbb{R}$ uniformly on any bounded set of X, and $\phi \in WPSAP_\omega(\mathbb{R}, X, \rho)$, then the function $f(\cdot, \phi(\cdot)) \in WPSAAP_\omega(\mathbb{R}, X, \rho)$.*

2.5 Stepanov-Like Asymptotically Bloch-Type Periodic Functions

In this section, we mainly introduce the notion of Stepanov-like (weighted) pseudo S-asymptotically Bloch type periodicity and establish some basic properties. In what follows, we let $p \in [1, \infty)$ and recall that $\mathcal{U}_{\mathrm{inv}} = \{\rho \in \mathcal{U}_\infty : \rho \text{ meets the hypothesis } H_\rho\}$.

Definition 2.13 (Henríquez *et al.*, 2008a; Henríquez, 2013). A measurable function $f : \mathbb{R} \to X$ is said to be p-Stepanov bounded if

$$\sup_{t \in \mathbb{R}} \int_t^{t+1} \|f(s)\|^p ds < \infty.$$

We denote the space of all p-Stepanov bounded functions by $BS^p(\mathbb{R}, X)$.

The space $BS^p(\mathbb{R}, X)$ with the norm $\|f\|_{S^p} = \sup_{t \in \mathbb{R}} \left(\int_t^{t+1} \|f(s)\|^p ds \right)^{\frac{1}{p}}$ is a Banach space. The notation $BS^p(\mathbb{R} \times X, X)$ represents all functions $f : \mathbb{R} \times X \to X$ which is p-Stepanov bounded uniformly in $x \in X$.

Definition 2.14. A function $f \in BS^p(\mathbb{R}, X)$ is said to be Stepanov-like S-asymptotically Bloch type periodic (or S^p-S-asymptotically (ω, k)-Bloch periodic) if for given $\omega, k \in \mathbb{R}$,

$$\lim_{|t| \to \infty} \int_t^{t+1} \left\| f(s + \omega) - e^{ik\omega} f(s) \right\|^p ds = 0, \ t \in \mathbb{R}.$$

We denote the space of all such functions by $S^p SABP_{\omega,k}(\mathbb{R}, X)$.

Definition 2.15. A function $f \in BS^p(\mathbb{R}, X)$ is said to be Stepanov-like pseudo S-asymptotically Bloch-type periodic (or S^p-pseudo S-asymptotically (ω, k)-Bloch periodic) if for given $\omega, k \in \mathbb{R}$,

$$\lim_{T \to \infty} \frac{1}{2T} \int_{-T}^T \left(\int_t^{t+1} \|f(s + \omega) - e^{ik\omega} f(s)\|^p ds \right)^{\frac{1}{p}} dt = 0, \quad t \in \mathbb{R}.$$

We denote the space of all such functions by $S^p PSABP_{\omega,k}(\mathbb{R}, X)$.

Definition 2.16. Let $\rho \in \mathcal{U}_\infty$. A function $f \in BS^p(\mathbb{R}, X)$ is said to be Stepanov-like weighted pseudo S-asymptotically (S^p-weighted pseudo S-asymptotically) Bloch (or (ω, k)) type periodic if for given $\omega, k \in \mathbb{R}$ and all $t \in \mathbb{R}$,

$$\lim_{T \to \infty} \frac{1}{\mu(T, \rho)} \int_{-T}^T \rho(t) \left(\int_t^{t+1} \|f(s + \omega) - e^{ik\omega} f(s)\|^p ds \right)^{\frac{1}{p}} dt = 0.$$

The set of all such functions is denoted by $S^p WPSABP_{\omega,k}(\mathbb{R}, X, \rho)$.

Remark 2.10. It is clear that Definitions 2.14–2.16 are reduced to concepts of Stepanov-like S-asymptotically ω-periodic functions (see Henríquez *et al.*, 2008a), Stepanov-like (weighted) pseudo S-asymptotically ω-periodic functions (see Xia *et al.*, 2017), respectively, if $\omega k = 2\pi$. Meanwhile, we can also obtain notions of Stepanov-like S-asymptotically ω-antiperiodic functions, Stepanov-like (weighted) pseudo S-asymptotically ω-anti-periodic functions, respectively, if $\omega k = \pi$.

Lemma 2.23. *The space $S^p SABP_{\omega,k}(\mathbb{R}, X)$ equipped with the norm $\|\cdot\|_{S^p}$ is a Banach space.*

Proof. We can get that $S^p SABP_{\omega,k}(\mathbb{R}, X)$ is a closed subspace of $BS^p(\mathbb{R}, X)$. Let $\{f_n\}_n \subseteq S^p SABP_{\omega,k}(\mathbb{R}, X)$ converge to f as $n \to \infty$. Then for any $\varepsilon > 0$, we can choose suitable constants $N > 0$ and $T_\varepsilon > 0$ such that

$$\int_t^{t+1} \left\| f_n(s+\omega) - e^{ik\omega} f_n(s) \right\|^p ds \le \left(\frac{\varepsilon}{3}\right)^p,$$

$$\int_t^{t+1} \left\| f_n(t) - f(t) \right\|^p ds \le \left(\frac{\varepsilon}{3}\right)^p,$$

for $n > N$ and $|t| > T_\varepsilon$. Thus,

$$\left(\int_t^{t+1} \left\| f(s+\omega) - e^{ik\omega} f(s) \right\|^p ds \right)^{\frac{1}{p}}$$

$$= \left(\int_t^{t+1} \left\| f(s+\omega) - f_n(s+\omega) + f_n(s+\omega) - e^{ik\omega} f_n(s) \right. \right.$$

$$\left. \left. + e^{ik\omega} f_n(s) - e^{ik\omega} f(s) \right\|^p ds \right)^{\frac{1}{p}}$$

$$\le \left(\int_t^{t+1} \left\| f(s+\omega) - f_n(s+\omega) \right\|^p ds \right)^{\frac{1}{p}}$$

$$+ \left(\int_t^{t+1} \left\| f_n(s+\omega) - e^{ik\omega} f_n(s) \right\|^p ds \right)^{\frac{1}{p}}$$

$$+ \left(\int_t^{t+1} \left\| e^{ik\omega} f_n(s) - e^{ik\omega} f(s) \right\|^p ds \right)^{\frac{1}{p}}$$

$$\leq \frac{\varepsilon}{3} + \frac{\varepsilon}{3} + \frac{\varepsilon}{3}$$

$$\leq \varepsilon,$$

which implies that the space $S^p SABP_{\omega,k}(\mathbb{R}, X)$ is a closed subspace of $BS^p(\mathbb{R}, X)$, so it is a Banach space with the $\|\cdot\|_{S^p}$ norm. $\qquad\square$

Lemma 2.24. *If $f \in SABP_{\omega,k}(\mathbb{R}, X)$, then $f \in S^p SABP_{\omega,k}(\mathbb{R}, X)$.*

Proof. Since f is a bounded continuous function, it is p-Stepanov bounded. Considering $f \in SABP_{\omega,k}(\mathbb{R}, X)$, we can get that for any $\varepsilon > 0$, there exists $T_\varepsilon > 0$ such that

$$\left\| f(t+\omega) - e^{ik\omega} f(t) \right\| \leq \varepsilon^{\frac{1}{p}}$$

for $|t| \geq T_\varepsilon$. This implies that

$$\int_t^{t+1} \left\| f(s+\omega) - e^{ik\omega} f(s) \right\|^p ds \leq \varepsilon$$

for $|t| \geq T_\varepsilon$, i.e., $f \in S^p SABP_{\omega,k}(\mathbb{R}, X)$. $\qquad\square$

Theorem 2.14. *Let $f \in BS^p(\mathbb{R} \times X, X)$ satisfy the following conditions:*

(D1) *For all $(t, x) \in \mathbb{R} \times X$, $f(t+\omega, x) = e^{ik\omega} f\left(t, e^{-ik\omega} x\right)$.*
(D2) *There exists a constant $L > 0$ such that for all $x, y \in X$ and $t \in \mathbb{R}$,*

$$\| f(t, x) - f(t, y) \| \leq L \| x - y \|.$$

Then for each $u(\cdot) \in S^p SABP_{\omega,k}(\mathbb{R}, X)$, we have

$$\nu(\cdot) = f(\cdot, u(\cdot)) \in S^p SABP_{\omega,k}(\mathbb{R}, X).$$

Proof. Since $u \in S^p SABP_{\omega,k}(\mathbb{R}, X)$, we can get that for any $\varepsilon > 0$, there exists $T_\varepsilon > 0$ such that

$$\int_t^{t+1} \| u(s+\omega) - e^{ik\omega} u(s) \|^p ds \leq \frac{\varepsilon}{L},$$

for every $|t| > T_\varepsilon$. Thus, we can deduce that

$$\int_t^{t+1} \| \nu(s+\omega) - e^{ik\omega} \nu(s) \|^p ds$$

$$= \int_t^{t+1} \| f(s+\omega, u(s+\omega)) - e^{ik\omega} f(s, u(s)) \|^p ds$$

$$= \int_t^{t+1} \| e^{ik\omega} f(s, e^{-ik\omega} u(s+\omega)) - e^{ik\omega} f(s, u(s)) \|^p ds$$

$$\leq L \int_t^{t+1} \|e^{-ik\omega}u(s+\omega) - u(s)\|^p ds$$

$$\leq \varepsilon$$

for any $|t| > T_\varepsilon$, which shows the assertion. $\qquad\square$

We introduce the following hypothesis:

(H1) For a strongly continuous function $\mathcal{T}(\cdot) : [0, \infty) \to \mathfrak{B}(X)$, there exists $\phi \in L^1(\mathbb{R}^+)$ such that $\|\mathcal{T}(t)\| \leq \phi(t)$ for all $t \in \mathbb{R}^+$, where $\phi : \mathbb{R}^+ \to \mathbb{R}^+$ is a nonincreasing function with

$$\phi_0 := \sum_{n=1}^{\infty} \phi(n) < \infty.$$

Theorem 2.15. *If (H1) holds and $f \in S^p SABP_{\omega,k}(\mathbb{R}, X)$, then*

$$u(t) = \int_{-\infty}^{t} \mathcal{T}(t-s)f(s)ds \in SABP_{\omega,k}(\mathbb{R}, X).$$

Proof. It is clear that

$$u(t) = \int_{-\infty}^{t} \mathcal{T}(t-s)f(s)ds = \int_0^{\infty} \mathcal{T}(s)f(t-s)ds$$

$$= \sum_{n=1}^{\infty} \int_{n-1}^{n} \mathcal{T}(s)f(t-s)ds.$$

Let

$$u_n(t) = \int_{n-1}^{n} \mathcal{T}(s)f(t-s)ds.$$

Note that $\phi_0 < \infty$, so the series $\sum_{n=1}^{\infty} \phi(n)$ is uniformly convergent on $\mathbb{R}$. On the other hand,

$$\|u_n(t)\| = \left\| \int_{n-1}^{n} \mathcal{T}(r)f(t-r)dr \right\| \leq \int_{t-n}^{t-n+1} \|\mathcal{T}(t-r)f(r)\| \, dr$$

$$\leq \phi(n-1) \left(\int_{t-n}^{t-n+1} \|f(r)\|^p dr \right)^{\frac{1}{p}}$$

$$\leq \phi(n-1)\|f\|_{S^p}.$$

It is concluded that the series $\sum_{n=1}^{\infty} u_n(t)$ is uniformly convergent on $\mathbb{R}$, consequently the uniform limit

$$u(t) := \int_{-\infty}^{t} \mathcal{T}(t-s)f(s)ds = \sum_{n=1}^{\infty} u_n(t) \in BC(\mathbb{R}, X).$$

Since $f \in S^p SABP_{\omega,k}(\mathbb{R}, X)$, for any $\varepsilon > 0$, we can choose suitable constants $T_\varepsilon > 0$ such that

$$\int_t^{t+1} \left\| f(s + \omega) - e^{ik\omega} f(s) \right\|^p ds \leq \varepsilon^p,$$

for every $|t| \geq T_\varepsilon$. Then

$$\left\| u_n(t + \omega) - e^{ik\omega} u_n(t) \right\|$$

$$= \left\| \int_{n-1}^n \mathcal{T}(s)[f(t + \omega - s) - e^{ik\omega} f(t - s)] ds \right\|$$

$$= \left\| \int_{t-n}^{t-n+1} \mathcal{T}(t - s)[f(s + \omega) - e^{ik\omega} f(s)] ds \right\|$$

$$\leq \phi(n - 1) \int_{t-n}^{t-n+1} \left\| f(s + \omega) - e^{ik\omega} f(s) \right\| ds$$

$$\leq \phi(n - 1) \left(\int_{t-n}^{t-n+1} \left\| f(s + \omega) - e^{ik\omega} f(s) \right\|^p ds \right)^{\frac{1}{p}}$$

$$\leq \phi(n - 1)\varepsilon,$$

for each $|t| \geq T_\varepsilon + n$, where $n \in [1, \infty)$, which implies that $u_n(t) \in SABP_{\omega,k}(\mathbb{R}, X)$. It is known that the space $SABP_{\omega,k}(\mathbb{R}, X)$ is a Banach space with the sup-norm, thus the uniform limit

$$u(t) := \int_{-\infty}^t \mathcal{T}(t - s) f(s) ds = \sum_{n=1}^\infty u_n(t) \in SABP_{\omega,k}(\mathbb{R}, X).$$

This ends the proof.   □

Lemma 2.25. *The space $S^p WPSABP_{\omega,k}(\mathbb{R}, X, \rho)$ is a Banach space with the norm $\| \cdot \|_{S^p}$ for each $\rho \in \mathcal{U}_\infty$.*

Proof. The proof can be conducted similarly to Lemma 2.23 by Definition 2.16.   □

Lemma 2.26. *Assume that $\rho \in \mathcal{U}_{inv}$. Then $WPSABP_{\omega,k}(\mathbb{R}, X, \rho) \subseteq S^p WPSABP_{\omega,k}(\mathbb{R}, X, \rho)$.*

Proof. Let $f \in WPSABP_{\omega,k}(\mathbb{R}, X, \rho)$, then we have $f(\cdot + s) \in WPSABP_{\omega,k}(\mathbb{R}, X, \rho)$ for $s \in [0, 1]$. Let $C > 0$ be a constant such that

$\|f(t+\omega) - e^{ik\omega}f(t)\| \leq C$ for all $t \in \mathbb{R}$. It follows from the Hölder inequality that

$$\frac{1}{\mu(T,\rho)} \int_{-T}^{T} \rho(t) \left(\int_{t}^{t+1} \|f(s+\omega) - e^{ik\omega}f(s)\|^p ds \right)^{\frac{1}{p}} dt$$

$$= \frac{1}{\mu(T,\rho)} \int_{-T}^{T} \rho^{\frac{1}{q}}(t) \rho^{\frac{1}{p}}(t) \left(\int_{t}^{t+1} \|f(s+\omega) - e^{ik\omega}f(s)\|^p ds \right)^{\frac{1}{p}} dt$$

$$\leq \frac{1}{\mu(T,\rho)} \left(\int_{-T}^{T} \rho(t) \int_{t}^{t+1} \|f(s+\omega) - e^{ik\omega}f(s)\|^p ds dt \right)^{\frac{1}{p}}$$

$$\times \left(\int_{-T}^{T} \rho(t) dt \right)^{\frac{1}{q}}$$

$$\leq C^{\frac{1}{q}} \left(\int_{0}^{1} \frac{1}{\mu(T,\rho)} \int_{-T}^{T} \rho(t) \|f(t+s+\omega) - e^{ik\omega}f(t+s)\| dt ds \right)^{\frac{1}{p}}$$

$$\to 0,$$

where $\frac{1}{p} + \frac{1}{q} = 1$, so $f \in S^p WPSABP_{\omega,k}(\mathbb{R}, X, \rho)$. $\qquad\square$

Lemma 2.27. *Let $\rho \in \mathcal{U}_\infty$ and $f \in BS^p(\mathbb{R}, X)$. Then the following assertions are equivalent:*

(a) $f \in S^p WPSABP_{\omega,k}(\mathbb{R}, X, \rho)$.
(b) *For each $\epsilon > 0$, $\lim_{T\to\infty} \frac{1}{\mu(T,\rho)} \int_{M_{T,\epsilon}(f)} \rho(t) dt = 0$, where*

$$M_{T,\epsilon}(f) = \left\{ t \in [-T,T] : \left(\int_{t}^{t+1} \|f(s+\omega) - e^{ik\omega}f(s)\|^p ds \right)^{\frac{1}{p}} \geq \epsilon \right\}.$$

Proof. (a) $\Rightarrow$ (b): Since $f \in S^p WPSABP_{\omega,k}(\mathbb{R}, X, \rho)$, we have

$$\lim_{T\to\infty} \frac{1}{\mu(T,\rho)} \int_{-T}^{T} \rho(t) \left(\int_{t}^{t+1} \|f(s+\omega) - e^{ik\omega}f(s)\|^p ds \right)^{\frac{1}{p}} dt = 0. \quad (2.5)$$

If the assertion (b) is not true, then there exists $\epsilon_0 > 0$ such that $\frac{1}{\mu(T,\rho)} \int_{M_{T,\epsilon_0}(f)} \rho(t) dt$ cannot converge to 0 as $T \to \infty$. Hence there exists

$\delta > 0$ such that for each n, $\frac{1}{\mu(T_n,\rho)} \int_{M_{T_n,\epsilon_0}(f)} \rho(t)dt \geq \delta$ for some $T_n > n$. It follows that

$$\frac{1}{\mu(T_n,\rho)} \int_{-T_n}^{T_n} \rho(t) \left(\int_t^{t+1} \|f(s+\omega) - e^{ik\omega}f(s)\|^p ds \right)^{\frac{1}{p}} dt$$

$$= \frac{1}{\mu(T_n,\rho)} \int_{M_{T_n,\epsilon_0}(f)} \rho(t) \left(\int_t^{t+1} \|f(s+\omega) - e^{ik\omega}f(s)\|^p ds \right)^{\frac{1}{p}} dt$$

$$+ \frac{1}{\mu(T_n,\rho)} \int_{[-T_n,T_n]\backslash M_{T_n,\epsilon_0}(f)} \rho(t) \left(\int_t^{t+1} \|f(s+\omega) - e^{ik\omega}f(s)\|^p ds \right)^{\frac{1}{p}} dt$$

$$\geq \frac{1}{\mu(T_n,\rho)} \int_{M_{T_n,\epsilon_0}(f)} \rho(t) \left(\int_t^{t+1} \|f(s+\omega) - e^{ik\omega}f(s)\|^p ds \right)^{\frac{1}{p}} dt$$

$$\geq \frac{\epsilon_0}{\mu(T_n,\rho)} \int_{M_{T_n,\epsilon_0}(f)} \rho(t)dt \geq \epsilon_0 \delta,$$

which is a contradiction to (2.5). So the assertion (b) is true if the assertion (a) holds.

(b) $\Rightarrow$ (a): Since $f \in BS^p(\mathbb{R}, X)$, we have $\|f\|_{S^p} < \infty$ and for any $\epsilon > 0$, there exists a constant $T_0 > 0$ such that for $T > T_0$, $\frac{1}{\mu(T,\rho)} \int_{M_{T,\epsilon}(f)} \rho(t)dt < \frac{\epsilon}{2\|f\|_{S^p}}$. Thus we have

$$\frac{1}{\mu(T,\rho)} \int_{-T}^{T} \rho(t) \left(\int_t^{t+1} \|f(s+\omega) - e^{ik\omega}f(s)\|^p ds \right)^{\frac{1}{p}} dt$$

$$= \frac{1}{\mu(T,\rho)} \int_{M_{T,\epsilon}(f)} \rho(t) \left(\int_t^{t+1} \|f(s+\omega) - e^{ik\omega}f(s)\|^p ds \right)^{\frac{1}{p}} dt$$

$$+ \frac{1}{\mu(T,\rho)} \int_{[-T,T]\backslash M_{T,\epsilon}(f)} \rho(t) \left(\int_t^{t+1} \|f(s+\omega) - e^{ik\omega}f(s)\|^p ds \right)^{\frac{1}{p}} dt$$

$$\leq 2\|f\|_{S^p} \frac{1}{\mu(T,\rho)} \int_{M_{T,\epsilon}(f)} \rho(t)dt + \epsilon \frac{1}{\mu(T,\rho)} \int_{[-T,T]\backslash M_{T,\epsilon}(f)} \rho(t)dt < 2\epsilon,$$

which implies that (2.5) holds, i.e., the assertion (a) is true. $\qquad\square$

Next, we establish some composition theorems for S^p-weighted pseudo S-asymptotically Bloch-type periodic function.

Theorem 2.16. *Let $\rho \in \mathcal{U}_\infty$. Assume that $f \in BS^p(\mathbb{R} \times X, X)$ satisfies (D1)–(D2). Then for each $u(\cdot) \in S^p WPSABP_{\omega,k}(\mathbb{R}, X, \rho)$, $\nu(\cdot) = f(\cdot, u(\cdot)) \in S^p WPSABP_{\omega,k}(\mathbb{R}, X, \rho)$.*

Proof. Since $u \in S^p WPSABP_{\omega,k}(\mathbb{R}, X, \rho)$, we can get that for any $\varepsilon > 0$, there exists a positive integer L_ε such that

$$\frac{1}{\mu(T,\rho)} \int_{-T}^{T} \rho(t) \left(\int_{t}^{t+1} \|u(s+\omega) - e^{ik\omega}u(s)\|^p ds \right)^{\frac{1}{p}} dt \leq \frac{\varepsilon}{L}$$

for every $T > L_\varepsilon$. Thus, it is deduced that

$$\frac{1}{\mu(T,\rho)} \int_{-T}^{T} \rho(t) \left(\int_{t}^{t+1} \|\nu(s+\omega) - e^{ik\omega}\nu(s)\|^p ds \right)^{\frac{1}{p}} dt$$

$$= \frac{1}{\mu(T,\rho)} \int_{-T}^{T} \rho(t) \left(\int_{t}^{t+1} \|f(s+\omega, u(s+\omega)) - e^{ik\omega}f(s, u(s))\|^p ds \right)^{\frac{1}{p}} dt$$

$$\leq \frac{1}{\mu(T,\rho)} \int_{-T}^{T} \rho(t) \left(\int_{t}^{t+1} \|f(s+\omega, u(s+\omega)) \right.$$

$$\left. - e^{ik\omega}f(s, e^{-ik\omega}u(s+\omega))\|^p ds \right)^{\frac{1}{p}} dt$$

$$+ \frac{1}{\mu(T,\rho)} \int_{-T}^{T} \rho(t) \left(\int_{t}^{t+1} \|e^{ik\omega}f(s, e^{-ik\omega}u(s+\omega)) \right.$$

$$\left. - e^{ik\omega}f(s, u(s))\|^p ds \right)^{\frac{1}{p}} dt$$

$$\leq \frac{1}{\mu(T,\rho)} \int_{-T}^{T} \rho(t) \left(\int_{t}^{t+1} \|e^{ik\omega}f(s, e^{-ik\omega}u(s+\omega)) \right.$$

$$\left. - e^{ik\omega}f(s, u(s))\|^p ds \right)^{\frac{1}{p}} dt$$

$$\leq \frac{1}{\mu(T,\rho)} \int_{-T}^{T} \rho(t) \left(\int_{t}^{t+1} \|f(s, e^{-ik\omega}u(s+\omega)) - f(s, u(s))\|^p ds \right)^{\frac{1}{p}} dt$$

$$\leq L \frac{1}{\mu(T,\rho)} \int_{-T}^{T} \rho(t) \left(\int_{t}^{t+1} \|e^{-ik\omega}u(s+\omega) - u(s)\|^p ds \right)^{\frac{1}{p}} dt \leq \varepsilon,$$

for any $T > L_\varepsilon$, which shows the claim. $\square$

Theorem 2.17. *Let $\rho \in \mathcal{U}_\infty$. Assume that $f \in BS^p(\mathbb{R} \times X, X)$ satisfies (D1) and*

(D3) There exists a function $L_f(\cdot) \in BS^r(\mathbb{R}, \mathbb{R}^+)$, $r \geq \max\{p, p/(p-1)\}$ such that

$$\|f(t,x) - f(t,y)\| \leq L_f(t)\|x - y\|, \quad x, y \in X, t \in \mathbb{R}.$$

Then for each $u(\cdot) \in S^p WPSABP_{\omega,k}(\mathbb{R}, X, \rho)$, there exists $q \in [1, p)$ such that $\nu(\cdot) = f(\cdot, u(\cdot)) \in S^q WPSABP_{\omega,k}(\mathbb{R}, X, \rho)$.

Proof. Since $r \geq p/p - 1$, there exists $q \in [1, p)$ such that $r = pq/(p-q)$. Let $p' = p/(p-q)$, $q' = p/q$, then $p' > 1$, $q' > 1$ and $1/p' + 1/q' = 1$. Owing to $u \in S^p WPSABP_{\omega,k}(\mathbb{R}, X, \rho)$, we can get that for any $\varepsilon > 0$, there exists a positive integer L_ε such that

$$\frac{1}{\mu(T, \rho)} \int_{-T}^{T} \rho(t) \left(\int_{t}^{t+1} \|u(s+\omega) - e^{ik\omega} u(s)\|^p ds \right)^{\frac{1}{p}} dt \leq \frac{\varepsilon}{\|L_f\|_{S^r}}$$

for every $T \geq L_\varepsilon$. It follows by the Minkowski inequality and Hölder inequality that for $T \geq L_\varepsilon$,

$$\left(\int_{t}^{t+1} \|\nu(s+\omega) - e^{ik\omega}\nu(s)\|^q ds \right)^{\frac{1}{q}}$$

$$= \left(\int_{t}^{t+1} \|f(s+\omega, u(s+\omega)) - e^{ik\omega} f(s, u(s))\|^q ds \right)^{\frac{1}{q}}$$

$$\leq \left(\int_{t}^{t+1} \|f(s+\omega, u(s+\omega)) - e^{ik\omega} f(s, e^{-ik\omega} u(s+\omega))\|^q ds \right)^{\frac{1}{q}}$$

$$+ \left(\int_{t}^{t+1} \|e^{ik\omega} f(s, e^{-ik\omega} u(s+\omega)) - e^{ik\omega} f(s, u(s))\|^q ds \right)^{\frac{1}{q}}$$

$$\leq \left(\int_{t}^{t+1} \|f(s+\omega, u(s+\omega)) - e^{ik\omega} f(s, e^{-ik\omega} u(s+\omega))\|^q ds \right)^{\frac{1}{q}}$$

$$+ \left(\int_{t}^{t+1} \|f(s, e^{-ik\omega} u(s+\omega)) - f(s, u(s))\|^q ds \right)^{\frac{1}{q}}$$

$$\leq \left(\int_{t}^{t+1} L_f^q(s) \|e^{-ik\omega} u(s+\omega) - u(s)\|^q ds \right)^{\frac{1}{q}}$$

$$\leq \left(\int_{t}^{t+1} L_f^q(s) \|u(s+\omega) - e^{ik\omega} u(s)\|^q ds \right)^{\frac{1}{q}}$$

$$\leq \left(\int_t^{t+1} L_f^{qp'}(s)ds \right)^{\frac{1}{qp'}} \left(\int_t^{t+1} \|u(s+\omega) - e^{ik\omega}u(s)\|^{qq'} ds \right)^{\frac{1}{qq'}}$$

$$\leq \left(\int_t^{t+1} L_f^r(s)ds \right)^{\frac{1}{r}} \left(\int_t^{t+1} \|u(s+\omega) - e^{ik\omega}u(s)\|^p ds \right)^{\frac{1}{p}}$$

$$\leq \|L_f\|_{S^r} \left(\int_t^{t+1} \|u(s+\omega) - e^{ik\omega}u(s)\|^p ds \right)^{\frac{1}{p}}.$$

Thus, we have

$$\frac{1}{\mu(T,\rho)} \int_{-T}^T \rho(t) \left(\int_t^{t+1} \|\nu(s+\omega) - e^{ik\omega}\nu(s)\|^q ds \right)^{\frac{1}{q}} dt$$

$$\leq \frac{\|L_f\|_{S^r}}{\mu(T,\rho)} \int_{-T}^T \rho(t) \left(\int_t^{t+1} \|u(s+\omega) - e^{ik\omega}u(s)\|^p ds \right)^{\frac{1}{p}} dt$$

$$\leq \varepsilon,$$

which implies that $\nu(\cdot) \in S^q WPSABP_{\omega,k}(\mathbb{R}, X, \rho)$. $\qquad\square$

Remark 2.11. The condition (D1) in Theorems 2.16 and 2.17 can be weakened by the following condition:

(D1′) For all $(t, x) \in \mathbb{R} \times X$,

$$\lim_{T\to\infty} \frac{1}{\mu(T,\rho)} \int_{-T}^T \rho(t) \left(\int_t^{t+1} \|f(s+\omega,x) - e^{ik\omega}f(s, e^{-ik\omega}x)\|^p ds \right)^{\frac{1}{p}} dt = 0$$

uniformly in $x \in X$.

In fact, the assertion of Theorem 2.16 is still true under conditions (D1′) and (D2). Similarly, the conclusion of Theorem 2.17 holds true if conditions (D1′) and (D3) are satisfied.

Theorem 2.18. *Let $\rho \in \mathcal{U}_\infty$. Assume further that $f \in BS^p(\mathbb{R} \times X, X)$ satisfies (D1′) and the following condition:* (D4) *For any $\epsilon > 0$, there exists $\delta > 0$ such that $\left(\int_0^1 \|f(t+s,x) - f(t+s,y)\|^p ds \right)^{1/p} < \epsilon$ for all $t \in \mathbb{R}$, $x, y \in X$ with $\|x - y\| < \delta$.*
Then for each $u(\cdot) \in S^p WPSABP_{\omega,k}(\mathbb{R}, X, \rho)$, $\nu(\cdot) = f(\cdot, u(\cdot)) \in S^p WPSABP_{\omega,k}(\mathbb{R}, X, \rho)$.

Proof. By (D4), for any $\epsilon > 0$, there exists $\delta > 0$ such that $\left(\int_t^{t+1} \| f(t + s, e^{-ik\omega} u(s + \omega)) - f(t + s, u(s)) \|^p ds \right)^{1/p} < \frac{\epsilon}{3}$ for all $t \in \mathbb{R}$ whenever $\| u(s + \omega) - e^{ik\omega} u(s) \| < \delta$ for all $s \in \mathbb{R}$. By (D1'), for above-selected $\epsilon > 0$, there exists $T_1 > 0$ such that $\frac{1}{\mu(T,\rho)} \int_{-T}^{T} \rho(t) \left(\int_t^{t+1} \| f(s + \omega, u(s + \omega)) - e^{ik\omega} f(s, e^{-ik\omega} u(s + \omega)) \|^p ds \right)^{\frac{1}{p}} dt < \frac{\epsilon}{3}$ for $T > T_1$. For each $u(\cdot) \in S^p WPSABP_{\omega,k}(\mathbb{R}, X, \rho)$ and above arbitrarily given $\epsilon > 0$, it follows from Lemma 2.27 that there exists $T_2 > 0$ such that for $T > T_2$, $\frac{1}{\mu(T,\rho)} \int_{M_{T,\delta}(u)} \rho(t) dt < \frac{\epsilon}{6\|\nu\|_{S^p}}$. Thus, for $T > T_0 = \max\{T_1, T_2\}$, we have for every $T > L_\varepsilon$. Thus, it is deduced that

$$
\frac{1}{\mu(T,\rho)} \int_{-T}^{T} \rho(t) \left(\int_t^{t+1} \| \nu(s + \omega) - e^{ik\omega} \nu(s) \|^p ds \right)^{\frac{1}{p}} dt
$$

$$
= \frac{1}{\mu(T,\rho)} \int_{-T}^{T} \rho(t) \left(\int_t^{t+1} \| f(s + \omega, u(s + \omega)) - e^{ik\omega} f(s, u(s)) \|^p ds \right)^{\frac{1}{p}} dt
$$

$$
\leq \frac{1}{\mu(T,\rho)} \int_{-T}^{T} \rho(t)
$$

$$
\times \left(\int_t^{t+1} \| f(s + \omega, u(s + \omega)) - e^{ik\omega} f(s, e^{-ik\omega} u(s + \omega)) \|^p ds \right)^{\frac{1}{p}} dt
$$

$$
+ \frac{1}{\mu(T,\rho)} \int_{-T}^{T} \rho(t)
$$

$$
\times \left(\int_t^{t+1} \| e^{ik\omega} f(s, e^{-ik\omega} u(s + \omega)) - e^{ik\omega} f(s, u(s)) \|^p ds \right)^{\frac{1}{p}} dt
$$

$$
\leq \frac{\epsilon}{3} + \frac{1}{\mu(T,\rho)} \int_{[-T,T] \setminus M_{T,\delta}(u)} \rho(t)
$$

$$
\times \left(\int_t^{t+1} \| e^{ik\omega} f(s, e^{-ik\omega} u(s + \omega)) - e^{ik\omega} f(s, u(s)) \|^p ds \right)^{\frac{1}{p}} dt
$$

$$
+ \frac{1}{\mu(T,\rho)} \int_{M_{T,\delta}(u)} \rho(t)
$$

$$
\times \left(\int_t^{t+1} \| e^{ik\omega} f(s, e^{-ik\omega} u(s + \omega)) - e^{ik\omega} f(s, u(s)) \|^p ds \right)^{\frac{1}{p}} dt
$$

$$\leq \frac{\epsilon}{3} + \frac{\epsilon}{3} \frac{1}{\mu(T,\rho)} \int_{[-T,T]\backslash M_{T,\delta}(u)} \rho(t)dt + 2\|\nu\|_{S^p} \frac{1}{\mu(T,\rho)}$$

$$\times \int_{M_{T,\delta}(u)} \rho(t)dt < \epsilon,$$

which implies that $\nu(\cdot) \in S^p \, WPSABP_{\omega,k}(\mathbb{R}, X, \rho)$. $\qquad\square$

In applications, we usually consider the weighted pseudo S-asymptotically (ω, k)-Bloch periodic function perturbed by a Stepanov-like force term. Thus we have the following composition property.

Corollary 2.10. *Let* $\rho \in \mathcal{U}_\infty$. *Assume that* $f \in BS^p(\mathbb{R} \times X, X)$ *verifies the following conditions*:

(D1″) *For all* $t \in \mathbb{R}$, $\lim_{T\to\infty} \frac{1}{\mu(T,\rho)} \int_{-T}^{T} \rho(t) \big(\int_{t}^{t+1} \|f(s + \omega, x) - e^{ik\omega} f(s, e^{-ik\omega}x)\|^p ds \big)^{\frac{1}{p}} dt = 0$ *uniformly in* $x \in \mathcal{Q}$ *any bounded subset of* X.

(D4′) *For any* $\epsilon > 0$, *there exists* $\delta > 0$ *such that* $\big(\int_{0}^{1} \|f(t+s, x) - f(t+s, y)\|^p ds \big)^{1/p} < \epsilon$ *for all* $t \in \mathbb{R}$, $x, y \in \mathcal{Q}$ *with* $\|x - y\| < \delta$.

Then for each $u(\cdot) \in WPSABP_{\omega,k}(\mathbb{R}, X, \rho)$, $\nu(\cdot) = f(\cdot, u(\cdot)) \in S^p \, WPSABP_{\omega,k}(\mathbb{R}, X, \rho)$.

Proof. For each $u(\cdot) \in WPSABP_{\omega,k}(\mathbb{R}, X, \rho)$, let $\mathcal{Q} = \overline{\{u(t) : t \in \mathbb{R}\}}$. Noting that Lemma 2.26, the remainder of the proof can be completed analogously to Theorem 2.18. $\qquad\square$

Theorem 2.19. *Let (H1) hold and* $f \in S^p \, WPSABP_{\omega,k}(\mathbb{R}, X, \rho)$ *for* $\rho \in \mathcal{U}_\infty$. *Then*

$$u(t) = \int_{-\infty}^{t} \mathcal{T}(t - s)f(s)ds \in WPSABP_{\omega,k}(\mathbb{R}, X, \rho).$$

Proof. The proof can be similarly conducted as Theorem 2.15, it is just to show that

$$u_n(t) \in WPSABP_{\omega,k}(\mathbb{R}, X, \rho).$$

Since $f \in S^p \, WPSABP_{\omega,k}(\mathbb{R}, X, \rho)$, we can get that there exists $L_\varepsilon > 0$, such that for $T \geq L_\varepsilon$

$$\frac{1}{\mu(T,\rho)} \int_{-T}^{T} \rho(t) \left(\int_{t}^{t+1} \|f(s + \omega) - e^{ik\omega} f(s)\|^p ds \right)^{\frac{1}{p}} dt < \varepsilon.$$

Then

$$\frac{1}{\mu(T,\rho)} \int_{-T}^{T} \rho(t) \left\| u_n(t+\omega) - e^{ik\omega} u_n(t) \right\| dt$$

$$= \frac{1}{\mu(T,\rho)} \int_{-T}^{T} \rho(t) \left\| \int_{n-1}^{n} \mathcal{T}(s)[f(t+\omega-s) - e^{ik\omega} f(t-s)]ds \right\| dt$$

$$= \frac{1}{\mu(T,\rho)} \int_{-T}^{T} \rho(t) \left\| \int_{t-n}^{t-n+1} \mathcal{T}(t-s)[f(s+\omega) - e^{ik\omega} f(s)]ds \right\| dt$$

$$\leq \frac{\phi(n-1)}{\mu(T,\rho)} \int_{-T}^{T} \rho(t) \int_{t-n}^{t-n+1} \left\| f(s+\omega) - e^{ik\omega} f(s) \right\| ds\,dt$$

$$\leq \frac{\phi(n-1)}{\mu(T,\rho)} \int_{-T}^{T} \rho(t) \left(\int_{t-n}^{t-n+1} \left\| f(s+\omega) - e^{ik\omega} f(s) \right\|^p ds \right)^{\frac{1}{p}} dt$$

$$\leq \phi(n-1)\varepsilon,$$

which implies that $u_n(t) \in WPSABP_{\omega,k}(\mathbb{R}, X, \rho)$, and it is known that $WPSABP_{\omega,k}(\mathbb{R}, X, \rho)$ is a Banach space with the sup-norm, thus the uniform limit

$$u(t) = \sum_{n=1}^{\infty} u_n(t) = \int_{-\infty}^{t} \mathcal{T}(t-s)f(s)ds \in WPSABP_{\omega,k}(\mathbb{R}, X, \rho).$$

This finishes the proof. □

Remark 2.12. If $\rho \equiv 1$, then Definition 2.16 is reduced to Definition 2.15. Thus, Lemmas 2.25–2.26, Theorems 2.16–2.19 still hold true for $S^p PSABP_{\omega,k}(\mathbb{R}, X)$ with $\rho \equiv 1$. For an application of the pseudo Stepanov-like S-asymptotically (ω, k)-Bloch periodic function, please refer to Chang and Wei (2022).

Chapter 3

Bloch-Type Periodic Solutions to Semilinear Integrodifferential Equations of Mixed Kernel

In this chapter, we mainly consider the following semilinear integrodifferential equation

$$u'(t) = Au(t) + \int_{-\infty}^t a(t-s)Au(s)ds + f(t, u(t)), \quad t \in \mathbb{R}, \qquad (3.1)$$

where $A : D(A) \subseteq X \to X$ is a closed linear operator defined on a Banach space X, the kernel function $a(t) := \alpha \frac{t^{\mu-1}}{\Gamma(\mu)} e^{-\beta t}$ with $\alpha \neq 0, \beta > 0, \mu \geq 1$, $f \in BC(\mathbb{R} \times X, X)$, and $\Gamma(\cdot)$ stands for the Gamma function.

Equation (3.1) can be used to describe the heat conduction in materials with memory (see for instance Meyers and Chawla, 2009; Ponce, 2020a). It is noticed that Eq. (3.1) has been treated in some special cases. For example, if $\mu = 1$, i.e., $a(t) := \alpha e^{-\beta t}$ then Eq. (3.1) is turned into the following form:

$$u'(t) = Au(t) + \alpha \int_{-\infty}^t e^{-\beta(t-s)} Au(s)ds + f(t, u(t)), \quad t \in \mathbb{R}. \qquad (3.2)$$

Lizama and Ponce (2011b) systematically studied the existence and uniqueness of bounded mild solutions, such as almost periodic (automorphic), pseudo almost periodic (automorphic), asymptotically almost periodic (automorphic), to Eq. (3.2). Chang *et al.* (2015b) also investigated some existence results on measure pseudo almost automorphic mild solutions to Eq. (3.2). If we consider the *limit* case $\beta = 0$ in Eq. (3.1), we obtain the following multi-term fractional differential equation

$$\partial_t^{\mu+1} u(t) = A\partial_t^\mu u(t) + \alpha Au(t) + F(t, u(t)), \quad t \in \mathbb{R}, \qquad (3.3)$$

where ∂_t^α denotes the Weyl fractional derivative (see next Chapter 4). Equations in the form of (3.3) have been widely studied, see for instance Alvarez-Pardo and Lizama (2015), Keyantuo *et al.* (2013), Ponce (2013a, 2013b) and references therein. Chang and Ponce (2018) proved the existence of a uniformly exponential stable resolvent family and investigated the existence and uniqueness of bounded mild solutions such as asymptotically periodic, asymptotically almost periodic (automorphic), to Eq. (3.1). Chang *et al.* (2018) considered the existence and uniqueness of (weighted pseudo) almost automorphic in distribution mild solutions to Eq. (3.1) perturbed by Lévy noise. Here, we continue to consider the existence and uniqueness of generalized Bloch type periodic mild solutions to Eq. (3.1) as an application. We can refer to Kostić (2019) for more results on almost periodic and almost automorphic solutions to integrodifferential equations in abstract spaces.

3.1   Uniform Exponential Stability of Solutions to a Volterra Equation

In this section, we recall some results (see Chang and Ponce, 2018) on the uniform exponential stability of solutions to the homogeneous abstract Volterra equation

$$\begin{cases} u'(t) = Au(t) + \displaystyle\int_0^t a(t-s)Au(s)ds, & t \geq 0, \\ u(0) = x, \end{cases} \tag{3.4}$$

where $a(t) := \alpha e^{-\beta t} \frac{t^{\mu-1}}{\Gamma(\mu)}$, $t > 0, \alpha \in \mathbb{R}$, $\beta > 0, \mu \geq 1$, A generates a C_0-semigroup on a Banach space X and $x \in X$. It is said that a solution of (3.4) is *uniformly exponentially stable* if there exist $\delta > 0$ and $C > 0$ such that for each $x \in D(A)$, the corresponding solution $u(t)$ satisfies

$$\|u(t)\| \leq Ce^{-\delta t}\|x\|, \quad t \geq 0.$$

As in Chen *et al.* (2009), to investigate the uniform exponential stability of solutions to (3.4), we introduce a matrix operator as follow. We define the operator $\mathcal{A}|_D$, on $\mathcal{X} := X \times \mathbb{M}$ by

$$\mathcal{A}|_D \begin{pmatrix} x \\ f \end{pmatrix} = \begin{pmatrix} A & \delta_0 \\ B & \frac{d}{ds} \end{pmatrix} \begin{pmatrix} x \\ f \end{pmatrix} \tag{3.5}$$

with domain $D(\mathcal{A}|_D) = D(A) \times \mathbb{M}$, where

$$Bx := a(\cdot)Ax, \delta_0 f = f(0), \quad \text{and} \quad \mathbb{M} := \{s^{\mu-1}e^{-\beta s}x : x \in X\}.$$

Observe that $\mathbb{M}$ is a closed subspace of $L^p(\mathbb{R}_+, X)$ for all $p \geq 1$. Now, we give some spectral properties of $\mathcal{A}|_D$.

Proposition 3.1. *Let* $a(t) := \alpha e^{-\beta t} \frac{t^{\mu-1}}{\Gamma(\mu)}$, *where* $\alpha \neq 0, \beta > 0$ *and* $\mu \geq 1$. *The following assertions hold:*

(1) $-\beta$ *is an eigenvalue of* $\mathcal{A}|_D$ *that is,* $-\beta \in \sigma_p(\mathcal{A}|_D)$ *if and only if* $0 \in \sigma_p(A)$.

(2) *If* $\dfrac{\alpha}{\beta^\mu} = -1$, *then* $0 \in \sigma_p(\mathcal{A}|_D)$.

(3) *If* $\dfrac{\alpha}{\beta^\mu} \neq -1$ *and* $\lambda \notin \{-\beta, (-\alpha)^{1/\mu} - \beta\}$, *then* $\lambda \in \sigma_p(\mathcal{A}|_D)$ *if and only if* $\lambda(\lambda + \beta)^\mu ((\lambda + \beta)^\mu + \alpha)^{-1} \in \sigma_p(A)$.

Proof. 1. Consider the equation

$$(\lambda - \mathcal{A}|_D)\begin{pmatrix} x \\ f \end{pmatrix} = 0,$$

which is equivalent to the initial value problem

$$\begin{cases} (\lambda - A)x - f(0) = 0 \\ -\alpha e^{-\beta s} \dfrac{s^{\mu-1}}{\Gamma(\mu)} Ax + \lambda f(s) - f'(s) = 0. \end{cases} \tag{3.6}$$

If $\lambda = -\beta$, then from (3.6) we obtain

$$\begin{cases} (-\beta - A)x - f(0) = 0 \\ f(s) = e^{-\beta s} f(0) - \alpha e^{-\beta s} \dfrac{s^\mu}{\Gamma(\mu+1)} Ax. \end{cases} \tag{3.7}$$

Thus, there exists a nonzero $f(s) \in \mathbb{M}$ if and only if $Ax = 0$ for some nonzero $x \in D(A)$. Therefore, $-\beta \in \sigma_p(\mathcal{A}|_D)$ if and only if $0 \in \sigma_p(A)$.

On the other hand, if $\lambda \neq -\beta$, then from (3.6) we get (using the method of variation of parameters) that $f(0) = \frac{\alpha}{(\lambda+\beta)^\mu} Ax$, and

$$f(s) = f(0)e^{\lambda s} - \frac{e^{\lambda s}}{\Gamma(\mu)} \frac{\alpha}{(\lambda+\beta)^\mu} \int_0^{(\lambda+\beta)s} v^{\mu-1} e^{-v} Ax \, dv. \tag{3.8}$$

The first equation in (3.6) yields

$$\frac{\alpha}{(\lambda+\beta)^\mu} Ax = (\lambda - A)x. \tag{3.9}$$

If $x = 0$, then from (3.9) we get $f(0) = 0$ and therefore from (3.8) we conclude that $f(s) = 0$ for all s. From (3.9) we conclude that $\lambda \in \sigma_p(\mathcal{A}|_D)$ if and only if there exists $x \neq 0$ such that

$$\left[\lambda - \left(1 + \frac{\alpha}{(\lambda + \beta)^\mu}\right) A\right] x = 0. \qquad (3.10)$$

2. If $\dfrac{\alpha}{\beta^\mu} = -1$, then by (3.10) (with $\lambda = 0$) we conclude $0 \in \sigma_p(\mathcal{A}|_D)$.

3. Suppose that $\dfrac{\alpha}{\beta^\mu} \neq -1$ and $\lambda \notin \{-\beta, (-\alpha)^{1/\mu} - \beta\}$. From (3.10) we have

$$\begin{aligned}
0 &= \left[\lambda - \left(1 + \frac{\alpha}{(\lambda + \beta)^\mu}\right) A\right] x \\
&= \left(1 + \frac{\alpha}{(\lambda + \beta)^\mu}\right)\left[\lambda \frac{(\lambda + \beta)^\mu}{(\lambda + \beta)^\mu + \alpha} - A\right] x,
\end{aligned}$$

and therefore, $\lambda \in \sigma_p(\mathcal{A}|_D)$ if and only if $\lambda \dfrac{(\lambda + \beta)^\mu}{(\lambda + \beta)^\mu + \alpha} \in \sigma_p(A)$. $\qquad\square$

Lemma 3.1. *Let* $a(t) := \alpha e^{-\beta t} \dfrac{t^{\mu-1}}{\Gamma(\mu)}$, *where* $\alpha \neq 0$, $\beta > 0$ *and* $\mu \geq 1$. *If* $\lambda \neq (-\alpha)^{1/\mu} - \beta$ *and* $\lambda(\lambda + \beta)^\mu((\lambda + \beta)^\mu + \alpha)^{-1} \in \rho(A)$, *then* $\lambda \in \rho(\mathcal{A}|_D)$.

Proof. Consider the eigen-equation

$$(\lambda - \mathcal{A}|_D)\begin{pmatrix} x \\ f \end{pmatrix} = \begin{pmatrix} y \\ g \end{pmatrix}, \qquad (3.11)$$

which is equivalent to the initial value problem

$$\begin{cases} (\lambda - A)x - f(0) = y, \\ -\alpha e^{-\beta s} \dfrac{s^{\mu-1}}{\Gamma(\mu)} Ax + \lambda f(s) - f'(s) = g(s). \end{cases} \qquad (3.12)$$

Multiplying by $e^{-\lambda s}$ the second equation in (3.12) and integrating from $s = 0$ to $s = \infty$ we get

$$\frac{(\lambda + \beta)^\mu + \alpha}{(\lambda + \beta)^\mu}\left[\frac{\lambda(\lambda + \beta)^\mu}{(\lambda + \beta)^\mu + \alpha} - A\right] x = \hat{g}(\lambda) + y.$$

By hypothesis, we conclude that

$$x = \frac{(\lambda + \beta)^\mu}{(\lambda + \beta)^\mu + \alpha}\left[\frac{\lambda(\lambda + \beta)^\mu}{(\lambda + \beta)^\mu + \alpha} - A\right]^{-1}[\hat{g}(\lambda) + y].$$

To find f in Eq. (3.11) we solve the initial value problem (3.12) and we obtain

$$f(s) = \left(-\int_0^s g(v)e^{-\lambda v}\,dv - \int_0^s \alpha e^{-\beta v}\frac{v^{\mu-1}}{\Gamma(\mu)}e^{-\lambda v} Ax\,dv\right) e^{\lambda s}.$$

Therefore, $\lambda \in \rho(\mathcal{A}|_D)$. $\qquad\square$

The following theorem is the main result in this section and it is an extension of Chen *et al.* (2009, Theorem 3.4).

Theorem 3.1. *Let $\alpha \neq 0$, $\beta > 0$, $\mu \geq 1$ be such that $\mathrm{Re}((-\alpha)^{1/\mu} - \beta) < 0$. Assume further that*

(a) *The operator A generates an immediately norm continuous C_0-semigroup on a Banach space X;*

(b) $\sup \{\mathrm{Re}\lambda, \lambda \in \mathbb{C} : \lambda(\lambda + \beta)^\mu((\lambda + \beta)^\mu + \alpha)^{-1} \in \sigma(A)\} < 0.$

Then, the solutions of the problem (3.4) are uniformly exponentially stable.

Proof. An easy computation shows that the space $\mathbb{M} := \{s^{\mu-1}e^{-\beta s}x : x \in X\}$ satisfies the hypothesis in Chen *et al.* (2009, Theorem 2.9). By Lemma 3.1 we obtain that

$$\sigma(\mathcal{A}|_D) \subset \{\lambda \in \mathbb{C} : \lambda(\lambda + \beta)^\mu((\lambda + \beta)^\mu + \alpha)^{-1} \in \sigma(A)\}$$
$$\bigcup \{(-\alpha)^{1/\mu} - \beta\}.$$

Therefore, the solutions to (3.4) are uniformly exponentially stable by Chen *et al.* (2009, Theorem 2.9). $\square$

Example 3.1. Take $X = \mathbb{C}$ and $A = -I$ (the identity operator) in the problem (3.4).

(1) If $\alpha = 1, \beta = 1, \mu = 2$, then $a(t) = te^{-t}$ and the solution to (3.4) is given by

$$u(t) = \left(\frac{1}{3}e^{-2t} + \frac{2}{3}e^{-t/2}\cos\left(\frac{\sqrt{3}}{2}t\right)\right)x,$$

and therefore, $\|u(t)\| \to 0$ as $t \to \infty$.

(2) If $\alpha = -1, \beta = 1, \mu = 2$, then $a(t) = -te^{-t}$ and the solution to (3.4) is given by

$$u(t) = \left(\frac{1}{3} + \frac{2}{3}e^{-3t/2}\cos\left(\frac{\sqrt{3}}{2}t\right)\right)x.$$

Therefore, $\|u(t)\| \to 1/3$ as $t \to \infty$.

(3) If $\alpha = 2, \beta = 2, \mu = 2$, then $a(t) = 2te^{-2t}$ and the solution to (3.4) is

$$u(t) = \left(\frac{1}{5}e^{-3t} + \frac{2}{5}e^{-t}(2\cos t + \sin t)\right)x.$$

Then $\|u(t)\| \to 0$ as $t \to \infty$.

(4) If $\alpha = -4, \beta = 2, \mu = 2$, then $a(t) = -4te^{-2t}$ and the solution to (3.4) is

$$u(t) = \left(\frac{1}{2} + \frac{1}{14} e^{-5t/2} \left(7 \cos \left(\frac{\sqrt{7}t}{2} \right) + \sqrt{7} \sin \left(\frac{\sqrt{7}t}{2} \right) \right) \right) x,$$

and therefore $\|u(t)\| \to 1/2$ as $t \to \infty$.

(5) If $\alpha = -1/2, \beta = 1, \mu = 3$, then $a(t) = -\frac{t^2}{4}e^{-t}$ and a numerical computation shows that the solution to (3.4) satisfies $\|u(t)\| \to 0$ as $t \to \infty$.

(6) If $\alpha = -2, \beta = 1, \mu = 3$, then $a(t) = -t^2e^{-t}$ and a numerical computation shows that the solution to (3.4) satisfies $\|u(t)\| \to \infty$ as $t \to \infty$.

Remark 3.1. An easy computation shows that the parameters α, β and μ in the cases $1, 3, 5$ in Example 3.1 satisfy the hypothesis in Theorem 3.1, whereas in the cases $2, 4, 6$ the hypothesis are not fulfilled. On the other hand, several examples in case $\mu = 1$ can be found in Chen *et al.* (2009).

Example 3.2. Consider the problem

$$\begin{cases} \dfrac{\partial u}{\partial t}(t, x) = \dfrac{\partial^2 u}{\partial x^2}(t, x) + \dfrac{1}{2} \displaystyle\int_0^t (t-s)e^{-\frac{(t-s)}{3}} \dfrac{\partial^2 u}{\partial x^2}(s, x)ds, \quad t \geq 0, \\ u(0, t) = u(\pi, t) = 0, \\ u(0) = u_0. \end{cases} \tag{3.13}$$

Let $X = L^2[0, \pi]$ and define $A := \frac{d^2}{dx^2}$, with domain

$$D(A) = \{g \in H^2[0, \pi] : g(0) = g(\pi) = 0\}.$$

Then (3.13) can be converted into the abstract form (3.4) with $\alpha = 1/2, \beta = 1/3$ and $\mu = 2$. It is well known that A generates an analytic (and hence immediately norm continuous) C_0-semigroup $\{T(t)\}_{t \geq 0}$ on X. Moreover, $\sigma(A) = \sigma_p(A) = \{-n^2 : n \in \mathbb{N}\}$. Since we must have $\lambda(\lambda + \beta)^\mu((\lambda + \beta)^\mu + \alpha)^{-1} \in \sigma(A)$, we need to solve the equation

$$\frac{\lambda(\lambda + \frac{1}{3})^2}{(\lambda + \frac{1}{3})^2 + \frac{1}{2}} = -n^2,$$

obtaining that the solutions are given by

$$\lambda_{n,1} = \frac{-(2 + 3n^2)}{9} - \frac{2^{\frac{2}{3}} a_n}{c_n} + \frac{1}{9\sqrt[3]{4}} c_n,$$

$$\lambda_{n,2} = \frac{-(2 + 3n^2)}{9} + \frac{(1 + \sqrt{3}i)a_n}{2^{\frac{1}{3}} c_n} - \frac{1}{18\sqrt[3]{4}}(1 - \sqrt{3}i)c_n,$$

$$\lambda_{n,3} = \frac{-(2 + 3n^2)}{9} + \frac{(1 - \sqrt{3}i)a_n}{2^{\frac{1}{3}} c_n} - \frac{1}{18\sqrt[3]{4}}(1 + \sqrt{3}i)c_n,$$

for all $n \geq 1$, where

$$a_n := -\frac{1}{9} + \frac{2}{3}n^2 - n^4,$$

$$c_n := \left(4 - 765n^2 + 108n^4 - 108n^6\right.$$

$$\left. + 27n\sqrt{-8 + 801n^2 - 216n^4 + 216n^6}\right)^{\frac{1}{3}}.$$

It is easy to see that

$$\sup\left\{\operatorname{Re}\lambda : \lambda\left(\lambda + \frac{1}{3}\right)^2\left(\left(\lambda + \frac{1}{3}\right)^2 + \frac{1}{2}\right)^{-1} \in \sigma(A)\right\} < 0.$$

Therefore, from Theorem 3.1 we conclude that the solution u to the problem (3.13) is uniformly exponentially stable.

3.2 Linear Integrodifferential Equation

The linear equation to Eq. (3.1) is such as

$$u'(t) = Au(t) + \int_{-\infty}^{t} a(t-s)Au(s)ds + f(t), \quad t \in \mathbb{R}. \tag{3.14}$$

The following proposition ensures the existence of a strongly continuous family of bounded linear operators, which commutes with A and satisfies certain resolvent equation. This class of strongly continuous families has been studied extensively in the literature of abstract Volterra equations, see for instance Prüss (1993) and references therein.

Proposition 3.2. *Let* $a(t) := \alpha\frac{t^{\mu-1}}{\Gamma(\mu)}e^{-\beta t}$ *where* $\alpha \neq 0, \beta > 0$ *and* $\mu \geq 1$ *and* $\operatorname{Re}((-\alpha)^{1/\mu} - \beta) < 0$. *Assume that*

(a) *The operator* A *generates an immediately norm continuous* C_0-*semigroup on a Banach space* X;
(b) $\sup\left\{\operatorname{Re}\lambda, \lambda \in \mathbb{C} : \lambda(\lambda + \beta)^\mu((\lambda + \beta)^\mu + \alpha)^{-1} \in \sigma(A)\right\} < 0.$

Then there exists a uniformly exponentially stable and strongly continuous family of operators $\{S(t)\}_{t\geq 0} \subseteq \mathcal{B}(X)$ *such that* $S(t)$ *commutes with* A, *that is,* $S(t)D(A) \subseteq D(A)$, $AS(t)x = S(t)Ax$ *for all* $x \in D(A)$, $t \geq 0$ *and*

$$S(t)x = x + \int_0^t b(t-s)AS(s)xds, \quad \text{for all} \quad x \in X, \ t \geq 0, \tag{3.15}$$

where $b(t) := 1 + \int_0^t a(s)ds, \ t \geq 0.$

Proof. For $t \geq 0$ and $x \in X$ define $S(t)x := u(t; x) = u(t)$, where $u(t)$ is the unique solution of Eq. (3.4). See Engel and Nagel (1999, Corollary 7.22, p. 449) for the existence of such solution and their strong continuity. We will see that $S(\cdot)x$ satisfies the resolvent equation (3.15). Since $S(t)x$ is the solution of (3.4), we have that $S(t)x$ is differentiable and satisfies

$$S'(t)x = AS(t)x + \int_0^t a(t-s)AS(s)x ds. \tag{3.16}$$

Integrating (3.16), we conclude from Fubini's theorem that,

$$
\begin{aligned}
S(t)x - x &= \int_0^t AS(s)x ds + \int_0^t \int_0^s a(s-\tau)AS(\tau)x d\tau ds \\
&= \int_0^t AS(s)x ds + \int_0^t \int_\tau^t a(s-\tau)AS(\tau)x ds d\tau \\
&= \int_0^t AS(s)x ds + \int_0^t \int_0^{t-\tau} a(v)AS(\tau)x dv d\tau \\
&= \int_0^t AS(s)x ds + \int_0^t (b(t-\tau) - 1)AS(\tau)x d\tau \\
&= \int_0^t b(t-\tau)AS(\tau)x d\tau.
\end{aligned}
$$

The commutativity of $S(t)$ with A follows in the same way that (Prüss, 1993, pp. 31–32). Finally, from Theorem 3.1 there exist $C, \delta > 0$ such that $\|S(t)\| \leq Ce^{-\delta t}$ for all $t \geq 0$, that is, $\{S(t)\}_{t \geq 0}$ is uniformly exponentially stable. $\qquad \square$

We recall that a function $u \in C^1(\mathbb{R}; X)$ is called a strong solution of (3.14) on $\mathbb{R}$ if $u \in C(\mathbb{R}; D(A))$ and (3.14) holds for all $t \in \mathbb{R}$. If $u(t) \in X$ instead of $u(t) \in D(A)$, and (3.14) holds for all $t \in \mathbb{R}$ we say that u is a mild solution of (3.14).

Let the notation $\mathcal{N}(X)$ to denote any one of spaces $BP_{\omega,k}(\mathbb{R}, X)$, $ABP_{\omega,k}(\mathbb{R}, X)$, $PBP_{\omega,k}(\mathbb{R}, X)$, $WPBP_{\omega,k}(\mathbb{R}, X, \rho)$ ($\rho \in \mathcal{U}_{\text{inv}}$), and $SABP_{\omega,k}(\mathbb{R}, X)$, $PSABP_{\omega,k}(\mathbb{R}, X)$, $WPSABP_{\omega,k}(\mathbb{R}, X, \rho)$ ($\rho \in \mathcal{U}_{\text{inv}}$) defined above. We have the following result.

Theorem 3.2. *Let* $a(t) := \alpha \frac{t^{\mu-1}}{\Gamma(\mu)} e^{-\beta t}$ *where* $\alpha \neq 0, \beta > 0$, $\mu \geq 1$ *and* $\mathrm{Re}((-\alpha)^{1/\mu} - \beta) < 0$. *Assume that conditions (a)–(b) in Proposition 3.2 hold. If* $f \in \mathcal{N}(X)$, *then the unique mild solution to (3.14) belongs to* $\mathcal{N}(X)$, *which is given by*

$$u(t) = \int_{-\infty}^{t} S(t-s)f(s)ds, \quad t \in \mathbb{R}, \tag{3.17}$$

where $\{S(t)\}_{t \geq 0}$ *is given in Proposition 3.2.*

Proof. By Proposition 3.2, the family $\{S(t)\}_{t \geq 0}$ is uniformly exponentially stable and therefore u is well defined. For a given $f \in \mathcal{N}(X)$, we have u defined by (3.17) belongs to $\mathcal{N}(X)$ through Lemma 2.1 and Theorems 2.1, 2.2, 2.7, 2.10, 2.13, respectively. Since S satisfies the resolvent equation

$$S(t)x = x + \int_{0}^{t} b(t-s)AS(s)xds, \quad x \in X,$$

where $b(t) = 1 + \int_{0}^{t} a(s)ds$, we have that b is differentiable and the above equation shows that for each $x \in X$, $S'(t)x$ exists and

$$S'(t)x = AS(t)x + \int_{0}^{t} a(t-s)AS(s)xds.$$

It remains to prove that u is a mild solution to (3.14). Since A is a closed operator, using the Fubini theorem, we have

$$u'(t) = S(0)f(t) + \int_{-\infty}^{t} S'(t-s)f(s)ds$$

$$= f(t) + \int_{-\infty}^{t} \left[AS(t-s)f(s) + \int_{0}^{t-s} a(t-s-\tau)AS(\tau)f(s)d\tau \right] ds$$

$$= f(t) + \int_{-\infty}^{t} AS(t-s)f(s)ds + \int_{-\infty}^{t} \int_{0}^{t-s} a(t-s-\tau)AS(\tau)f(s)d\tau ds$$

$$= f(t) + Au(t) + \int_{-\infty}^{t} \int_{s}^{t} a(t-v)AS(v-s)f(s)dvds$$

$$= f(t) + Au(t) + \int_{-\infty}^{t} \int_{-\infty}^{v} a(t-v)AS(v-s)f(s)dsdv$$

$$= f(t) + Au(t) + \int_{-\infty}^{t} a(t-v) \int_{-\infty}^{v} AS(v-s)f(s)dsdv$$

$$= f(t) + Au(t) + \int_{-\infty}^{t} a(t-v)Au(v)dv.$$

This ends the proof. $\qquad\qquad\qquad\qquad\qquad\qquad\qquad\qquad\qquad\square$

In case of Hilbert spaces, we can refer to a result of You (1992) (see also Lemma 1.9) which characterizes norm continuity of C_0-semigroups and obtain the following result.

Corollary 3.1. *Let A be the generator of a C_0-semigroup on a Hilbert space H. Let $s(A) := \sup\{\mathrm{Re}\lambda : \lambda \in \sigma(A)\}$ be the spectral bound of A. Let $\alpha \neq 0, \beta, \mu > 0$ with $\mathrm{Re}((-\alpha)^{1/\mu} - \beta) < 0$. Assume further that*

(a') $\lim_{\mu \in \mathbb{R}, |\mu| \to \infty} \|(\mu_0 + i\mu - A)^{-1}\| = 0$ *for some $\mu_0 > s(A)$;*
(b') $\sup\left\{\mathrm{Re}\lambda : \lambda(\lambda + \beta)^{\mu}((\lambda + \beta)^{\mu} + \alpha)^{-1} \in \sigma(A)\right\} < 0.$

If $f \in \mathcal{N}(H)$, then the unique mild solution to (3.14) belongs to $\mathcal{N}(H)$.

Remark 3.2. In the case $A = \varrho I$, $\varrho \in \mathbb{C}$ we obtain from (3.15), using Laplace transform, that for each $x \in X$:

$$\hat{S}_\varrho(\lambda)x = \frac{(\lambda + \beta)^{\mu}x}{\lambda(\lambda + \beta)^{\mu} - \varrho(\lambda + \beta)^{\mu} - \alpha\varrho}$$

$$= \frac{(\lambda + \beta)^{\mu}x}{(\lambda + \beta)^{\mu+1} - (\lambda + \beta)^{\mu}(\varrho + \beta) - \alpha\varrho}. \tag{3.18}$$

If $\beta + \varrho = 0$, then

$$\hat{S}_\varrho(\lambda)x = \frac{(\lambda + \beta)^{\mu}x}{(\lambda + \beta)^{\mu+1} - \alpha\varrho} = \frac{(\lambda + \beta)^{\mu+1-1}x}{(\lambda + \beta)^{\mu+1} - \alpha\varrho}$$

and therefore,

$$S_\varrho(t)x = e^{-\beta t} E_{\mu+1,1}(\alpha\varrho t^{\mu+1})x, \tag{3.19}$$

where $E_{a,b}(z)$ denotes the Mittag–Leffler function.

Recall that the Mittag–Leffler function (see, e.g., Agrawal *et al.*, 2007) is defined as follows:

$$E_{a,b}(z) := \sum_{k=0}^{\infty} \frac{z^k}{\Gamma(ak+b)} = \frac{1}{2\pi i} \int_{Ha} e^{\nu} \frac{\nu^{a-b}}{\nu^a - z} d\nu, \quad a, b > 0, \ z \in \mathbb{C},$$

where Ha is a Hankel path, i.e., a contour which starts and ends at $-\infty$ and encircles the disc $|\nu| \leq |z|^{1/a}$ counterclockwise. The Laplace transform of the Mittag–Leffler function is given by:

$$\mathfrak{L}\big(t^{b-1}E_{a,b}(-\varrho t^a)\big)(\lambda) = \frac{\lambda^{a-b}}{\lambda^a + \varrho}, \quad \varrho \in \mathbb{C}, \ \mathrm{Re}\lambda > |\varrho|^{1/a}.$$

The following result is concerned with the case $0 < \mu < 1$.

Theorem 3.3. *Let $A := \varrho I$ where $\varrho \in \mathbb{R}$ is given. Suppose that $\beta = -\varrho$, $\text{Re}((-\alpha)^{1/\mu} - \beta) < 0$ and $\alpha\varrho < 0$ for $0 < \mu < 1$. Let $f \in \mathcal{N}(X)$. Consider the equation*

$$u'(t) = \varrho u(t) + \varrho\alpha \int_{-\infty}^{t} \frac{(t-s)^{\mu-1}}{\Gamma(\mu)} e^{-\beta(t-s)} u(s) ds$$

$$+ f(t), \quad t \in \mathbb{R}. \tag{3.20}$$

Then Eq. (3.20) has a unique mild solution u which belongs to $\mathcal{N}(X)$ and is given by

$$u(t) = \int_{-\infty}^{t} S_\varrho(t-s) f(s) ds, \quad t \in \mathbb{R}, \tag{3.21}$$

where $\{S_\varrho(t)\}_{t \geq 0}$ is defined by (3.19).

Proof. Since $A = \varrho I$ generates an immediately norm continuous C_0-semigroup and $\sigma(A) = \{\varrho\}$, we have that

$$\lambda(\lambda + \beta)^\mu((\lambda + \beta)^\mu + \alpha)^{-1} \in \sigma(A)$$

if and only if

$$\lambda(\lambda + \beta)^\mu - \varrho(\lambda + \beta)^\mu - \alpha\varrho = 0.$$

We claim that $S_\varrho(t)$ is integrable. In fact, by Cuesta (2007, Theorem 1) there exists a constant $C_1 > 0$ depending only on α, ϱ and μ such that for all $t \geq 0, x \in X$

$$\left| t^\mu E_{\mu+1,1}(\alpha\varrho t^{\mu+1}) \right| = \left| t^{(\mu+1)-1} E_{\mu+1,1}(\alpha\varrho t^{\mu+1}) \right| \leq \frac{C_1}{1 + |\alpha\varrho| t^{\mu+1}}.$$

Therefore,

$$\int_0^\infty \|S_\varrho(t)\| dt \leq \int_0^\infty |e^{-\beta t} t^{-\mu} t^{(\mu+1)-1} E_{\mu+1,1}(\alpha\varrho t^{\mu+1})| dt$$

$$\leq C_1 \int_0^\infty \frac{e^{-\beta t} t^{-\mu}}{1 + |\alpha\varrho| t^{\mu+1}} dt$$

$$\leq C_1 \int_0^\infty e^{-\beta t} t^{-\mu} dt = \frac{C_1}{\beta^{1-\mu}} \Gamma(1 - \mu).$$

We conclude that $S_\varrho(t)$ is integrable, proving the claim. Therefore, from Theorem 3.2, Lemma 2.1 and Theorems 2.1, 2.2, 2.7, 2.10, 2.13, there exists a unique mild solution of Eq. (3.20) which belongs to $\mathcal{N}(X)$ and is explicitly given by (3.21). $\square$

Example 3.3. Let $\varrho = -1, \mu = 1/2, \alpha = 1/2, \beta = 1$. Hence, by Theorem 3.3, for any $f \in \mathcal{N}(X)$ there exists a unique mild solution $u \in \mathcal{N}(X)$ to the equation

$$u'(t) = -u(t) - \int_{-\infty}^t \frac{(t-s)^{-1/2}}{2\Gamma(1/2)} e^{-(t-s)} u(s)ds + f(t), \quad t \in \mathbb{R}, \quad (3.22)$$

given by

$$u(t) = \int_{-\infty}^t e^{-(t-s)} E_{3/2,1}\left(-\frac{(t-s)^{3/2}}{2}\right) f(s)ds, \ t \in \mathbb{R},$$

since $S_{-1}(t) = e^{-t} E_{3/2,1}\left(-\frac{t^{3/2}}{2}\right)$.

3.3 Semilinear Integrodifferential Equation

In this section, we investigate the existence and uniqueness of mild solutions in $\mathcal{N}(X)$ for Eq. (3.1).

Definition 3.1. A function $u : \mathbb{R} \to X$ is said to be a mild solution to Eq. (3.1) if

$$u(t) = \int_{-\infty}^t S(t-s)f(s, u(s))ds,$$

for all $t \in \mathbb{R}$, where $\{S(t)\}_{t \geq 0}$ is given in Proposition 3.2.

Theorem 3.4. Let $a(t) := \alpha \frac{t^{\mu-1}}{\Gamma(\mu)} e^{-\beta t}$ with $\alpha \neq 0, \beta > 0, \mu \geq 1$ and $\text{Re}((-\alpha)^{1/\mu} - \beta) < 0$. Assume that conditions (a)–(b) in Proposition 3.2 hold. Suppose further that $f = g + h \in BC(\mathbb{R} \times X, X)$ with g satisfying the condition (A1) in Section 2.3 and $h \in \mathscr{E}(\mathbb{R} \times X, X)$. If there exists a constant $0 < l < \delta/C$, where C, δ are given in Proposition 3.2, such that

$$\|f(t, x) - f(t, y)\| \leq l\|x - y\|,$$

for all $x, y \in X$ and $t \in \mathbb{R}$, then Eq. (3.1) has a unique mild solution $u \in PBP_{\omega,k}(\mathbb{R}, X)$.

Proof. Define the operator $\Upsilon : PBP_{\omega,k}(\mathbb{R}, X) \to PBP_{\omega,k}(\mathbb{R}, X)$ by

$$(\Upsilon u)(t) := \int_{-\infty}^t S(t-s)f(s, u(s))\, ds, t \in \mathbb{R}, \quad (3.23)$$

where $\{S(t)\}_{t \geq 0}$ is given in Proposition 3.2. For each u belonging to $PBP_{\omega,k}(\mathbb{R}, X)$, by (1) in Corollary 2.2, the function $s \mapsto f(s, u(s))$ belongs to $PBP_{\omega,k}(\mathbb{R}, X)$. It follows from Theorem 3.2 that $\Upsilon u \in PBP_{\omega,k}(\mathbb{R}, X)$, which implies that Υ is well-defined. By Proposition 3.2, there exist $\delta > 0, C > 0$ such that $\|S(t)\| \le Ce^{-\delta t}$ for all $t \ge 0$. For $u_1, u_2 \in PBP_{\omega,k}(\mathbb{R}, X)$ and $t \in \mathbb{R}$, we have

$$\|(\Upsilon u_1)(t) - (\Upsilon u_2)(t)\|$$

$$\le \int_{-\infty}^{t} \|S(t-s)[f(s, u_1(s)) - f(s, u_2(s))]\| ds$$

$$\le \int_{-\infty}^{t} l\|S(t-s)\| \cdot \|u_1(s) - u_2(s)\| ds$$

$$\le l\|u_1 - u_2\|_\infty \int_0^\infty \|S(r)\| dr$$

$$\le \frac{lC}{\delta} \|u_1 - u_2\|_\infty.$$

Thus

$$\|\Upsilon u_1 - \Upsilon u_2\|_\infty \le \frac{lC}{\delta} \|u_1 - u_2\|_\infty,$$

which shows that Υ is a contraction owing to the assumption $0 < l < \delta/C$. By the Banach fixed point theorem there exists a unique $u \in PBP_{\omega,k}(\mathbb{R}, X)$ such that $\Upsilon u = u$. The proof is ended. $\square$

Theorem 3.5. *Let* $\rho \in \mathcal{U}_{\text{inv}}$, $a(t) := \alpha \frac{t^{\mu-1}}{\Gamma(\mu)} e^{-\beta t}$ *with* $\alpha \ne 0, \beta > 0, \mu \ge 1$ *and* $\text{Re}((-\alpha)^{1/\mu} - \beta) < 0$. *Assume that conditions (a)–(b) in Proposition 3.2 hold. Suppose further that* $f = g + h \in BC(\mathbb{R} \times X, X)$ *with* g *satisfying the condition (A1) in Section 2.3 and* $h \in \mathcal{E}(\mathbb{R} \times X, X, \rho)$. *If there exists a constant* $0 < \tilde{l} < \delta/C$, *where* C, δ *are given in Proposition 3.2, such that*

$$\|f(t, x) - f(t, y)\| \le \tilde{l} \|x - y\|,$$

for all $x, y \in X$ *and* $t \in \mathbb{R}$, *then Eq. (3.1) has a unique mild solution* $u \in WPBP_{\omega,k}(\mathbb{R}, X, \rho)$.

Proof. The proof is analogous to Theorem 3.4 by Corollary 2.2(2) and Theorem 3.2. $\square$

Theorem 3.6. *Let* $a(t) := \alpha \frac{t^{\mu-1}}{\Gamma(\mu)} e^{-\beta t}$ *with* $\alpha \ne 0, \beta > 0, \mu \ge 1$ *and* $\text{Re}((-\alpha)^{1/\mu} - \beta) < 0$. *Assume that conditions (a)–(b) in Proposition 3.2 hold. Suppose further that* $f = g + h \in BC(\mathbb{R} \times X, X)$ *with* g *satisfying the condition (A1) in Section 2.3 and* $h \in \mathcal{E}(\mathbb{R} \times X, X)$. *If the condition (C1) in Corollary 2.3 is satisfied, then Eq. (3.1) has a unique mild solution* $u \in PBP_{\omega,k}(\mathbb{R}, X)$.

Proof. Let the operator Υ be defined as (3.23). For each $u \in PBP_{\omega,k}(\mathbb{R}, X)$, according to Corollary 2.3, Proposition 3.2 and Theorem 3.2 we have $\Upsilon u \in PBP_{\omega,k}(\mathbb{R}, X)$. Thus, Υ is well-defined.

For $L(\cdot) \in L^p(\mathbb{R}, \mathbb{R}_+)$ with $1 < p < \infty$, let $\tau(t) = \int_{-\infty}^{t} L^p(s)ds$. Define an equivalent norm over $PBP_{\omega,k}(\mathbb{R}, X)$ by

$$\|u\|_\tau = \sup_{t \in \mathbb{R}}\{e^{-\theta\tau(t)}\|u\|_\infty\}, \quad u \in PBP_{\omega,k}(\mathbb{R}, X),$$

where $\theta > 0$, is a constant sufficiently large. Now for each $u, v \in PBP_{\omega,k}(\mathbb{R}, X)$, we have

$$\|(\Upsilon u)(t) - (\Upsilon v)(t)\|$$

$$\leq \int_{-\infty}^{t} \|S(t-s)[f(s, u(s)) - f(s, v(s))]\|ds$$

$$\leq C \int_{-\infty}^{t} e^{-\delta(t-s)} L(s)\|u(s) - v(s)\|ds$$

$$\leq C \int_{-\infty}^{t} e^{-\delta(t-s)} L(s)e^{\theta\tau(s)}\|u-v\|_\tau ds$$

$$\leq C \left[\int_{-\infty}^{t} e^{\theta p\tau(s)} L^p(s)ds\right]^{\frac{1}{p}} \left[\int_{-\infty}^{t} e^{-\delta\frac{p(t-s)}{p-1}}ds\right]^{\frac{p-1}{p}} \|u-v\|_\tau$$

$$\leq C \left(\delta\frac{p}{p-1}\right)^{\frac{1-p}{p}} \left[\int_{-\infty}^{t} e^{\theta p\tau(s)}d\tau(s)\right]^{\frac{1}{p}} \|u-v\|_\tau$$

$$\leq C \left(\delta\frac{p}{p-1}\right)^{\frac{1-p}{p}} (p\theta)^{-\frac{1}{p}} e^{\theta\tau(t)}\|u-v\|_\tau.$$

Consequently,

$$\|\Upsilon u - \Upsilon v\|_\tau \leq C \left(\delta\frac{p}{p-1}\right)^{\frac{1-p}{p}} (p\theta)^{-\frac{1}{p}}\|u-v\|_\tau.$$

We can see that the operator Υ is a contraction for a sufficiently large θ.

For $L(\cdot) \in L^1(\mathbb{R}, \mathbb{R}_+)$, we have

$$\|(\Upsilon u)(t) - (\Upsilon v)(t)\| \leq \int_{-\infty}^{t} \|S(t-s)[f(s, u(s)) - f(s, v(s))]\|ds$$

$$\leq C \int_{-\infty}^{t} L(s)\|u(s) - v(s)\|ds$$

$$\leq C\|u-v\|_\infty \int_{-\infty}^{t} L(s)ds.$$

In the general, we have

$$\|(\Upsilon^n u)(t) - (\Upsilon^n v)(t)\|$$

$$\leq \frac{C^n}{(n-1)!}\|u-v\|_\infty \left[\int_{-\infty}^t L(s)\left(\int_{-\infty}^s L(\sigma)d\sigma\right)^{n-1} ds\right]$$

$$\leq \frac{C^n}{n!}\|u-v\|_\infty \left(\int_{-\infty}^t L(s)ds\right)^n.$$

Thus,

$$\|\Upsilon^n u - \Upsilon^n v\|_\infty \leq \frac{(C\|L\|_{L^1(\mathbb{R},\mathbb{R}_+)})^n}{n!}\|u-v\|_\infty.$$

Since $\frac{(C\|L\|_{L^1(\mathbb{R},\mathbb{R}_+)})^n}{n!} < 1$ for n sufficiently large, Υ is also a contraction.

From above arguments and the Banach fixed point theorem (see Lemma 1.21 and Corollary 1.1), there exists a unique $u \in PBP_{\omega,k}(\mathbb{R}, X)$ such that $\Upsilon u = u$. The proof is finished.   □

Theorem 3.7. *Let* $\rho \in \mathcal{U}_{\text{inv}}$, $a(t) := \alpha\frac{t^{\mu-1}}{\Gamma(\mu)}e^{-\beta t}$ *with* $\alpha \neq 0, \beta > 0$, $\mu \geq 1$ *and* $\text{Re}((-\alpha)^{1/\mu} - \beta) < 0$. *Assume that conditions* (a)–(b) *in Proposition 3.2 hold. Suppose further that* $f = g + h \in BC(\mathbb{R} \times X, X)$ *with* g *satisfying the condition* (A1) *in Section 2.3 and* $h \in \mathscr{E}(\mathbb{R} \times X, X, \rho)$. *If there exists a function* $L(\cdot) \in L^1(\mathbb{R}, \mathbb{R}_+)$ *such that the condition* (C2) *in Corollary 2.4 holds, then Eq.* (3.1) *has a unique mild solution* $u \in WPBP_{\omega,k}(\mathbb{R}, X, \rho)$.

Proof. For each $u \in WPBP_{\omega,k}(\mathbb{R}, X, \rho)$, define the operator Υ as (3.23). We can show that the operator Υ is well defined via Corollary 2.4, Theorem 3.2. Taking into account $L(\cdot) \in L^1(\mathbb{R}, \mathbb{R}_+)$, we can show Υ is a contraction analogously to the proof of Theorem 3.6 in the case $p = 1$. Thus, there exists a unique $u \in WPBP_{\omega,k}(\mathbb{R}, X, \rho)$ such that $\Upsilon u = u$ by the Banach fixed point theorem. This end the proof.   □

Theorem 3.8. *Let* $a(t) := \alpha\frac{t^{\mu-1}}{\Gamma(\mu)}e^{-\beta t}$ *with* $\alpha \neq 0, \beta > 0$, $\mu \geq 1$ *and* $\text{Re}((-\alpha)^{1/\mu} - \beta) < 0$. *Assume that conditions* (a)–(b) *in Proposition 3.2 hold. Suppose further that* $f \in BC(\mathbb{R} \times X, X)$ *satisfies* (T1)–(T2) *in Theorem 2.5 with* $0 < L < \delta/C$. *Then Eq.* (3.1) *admits a unique mild solution* $u \in SABP_{\omega,k}(\mathbb{R}, X)$.

Proof. For each $u \in SABP_{\omega,k}(\mathbb{R}, X)$, define the operator Υ as (3.23). According to Theorems 2.5 and 3.2, Υu also belongs to $SABP_{\omega,k}(\mathbb{R}, X)$ whenever $u \in SABP_{\omega,k}(\mathbb{R}, X)$. We can show that Υ is a contraction analogously to the proof of Theorem 3.4.   □

From Theorems 3.3, 3.4, and 3.8, we have the following results.

Corollary 3.2. *Let $A := \varrho I$ where $\varrho \in \mathbb{R}$ is given. Assume that $\beta = -\varrho$, $\mathrm{Re}((-\alpha)^{1/\mu} - \beta) < 0$ and $\alpha\varrho < 0$ for $0 < \mu < 1$. Consider the semilinear equation*

$$u'(t) = \varrho u(t) + \varrho\alpha \int_{-\infty}^{t} \frac{(t-s)^{\mu-1}}{\Gamma(\mu)} e^{-\beta(t-s)} u(s)ds$$

$$+ f(t, u(t)), \ t \in \mathbb{R}. \tag{3.24}$$

Let $f = g + h \in BC(\mathbb{R} \times X, X)$ satisfy $\|f(t,u) - f(t,v)\| \leq L\|u - v\|$, for all $t \in \mathbb{R}$ and $u, v \in X$, where $0 < L < \frac{\beta^{1-\mu}}{C_1\Gamma(1-\mu)}$, g verifies (A1) in Section 2.3, $h \in \mathscr{E}(\mathbb{R} \times X, X)$, and C_1 is given in the proof of Theorem 3.3. Then Eq. (3.24) has a unique solution $u \in PBP_{\omega,k}(\mathbb{R}, X)$ which is given by

$$u(t) = \int_{-\infty}^{t} S_\varrho(t-s)f(s, u(s))ds, \ t \in \mathbb{R},$$

where $\{S_\varrho(t)\}_{t \geq 0}$ is defined in (3.19).

Corollary 3.3. *Let $A := \varrho I$ where $\varrho \in \mathbb{R}$ is given. Suppose that $\beta = -\varrho$, $\mathrm{Re}((-\alpha)^{1/\mu} - \beta) < 0$ and $\alpha\varrho < 0$ for $0 < \mu < 1$. Let $f \in BC(\mathbb{R} \times X, X)$ satisfy (T1)–(T2) in Theorem 2.5, where $0 < L < \frac{\beta^{1-\mu}}{C_1\Gamma(1-\mu)}$, and C_1 is given in the proof of Theorem 3.3. Then Eq. (3.24) has a unique solution $u \in SABP_{\omega,k}(\mathbb{R}, X)$ which is given by*

$$u(t) = \int_{-\infty}^{t} S_\varrho(t-s)f(s, u(s))ds, \ t \in \mathbb{R},$$

where $\{S_\varrho(t)\}_{t \geq 0}$ is defined in (3.19).

Remark 3.3. Combined with Theorems 2.8, 2.11, and 3.2, we can also establish the existence and uniqueness of (weighted) pseudo S-asymptotically (ω, k)-Bloch periodic mild solutions to Eq. (3.1) under the condition that $f \in BC(\mathbb{R} \times X, X)$ satisfying (T1)–(T2) in Theorem 2.5 with $0 < L < \delta/C$.

Remark 3.4. For a Hilbert space H, according to Corollary 3.1, we can obtain that Eq. (3.1) admits a unique mild solution $u \in \mathcal{N}(H)$ via Theorems 3.4–3.8 and Remark 3.3.

Example 3.4. Let $X := L^2[0, \pi]$ and $\zeta \in [0, \pi], t \in \mathbb{R}$. Consider the following equation

$$\begin{cases} \dfrac{\partial x}{\partial t}(t, \zeta) = \dfrac{\partial^2 x}{\partial \zeta^2}(t, \zeta) + \dfrac{1}{2} \displaystyle\int_{-\infty}^{t} \dfrac{(t-s)}{\Gamma(2)} e^{-(t-s)} \dfrac{\partial^2 x}{\partial \zeta^2}(s, \zeta) ds \\ \qquad + g(t, x(t, \zeta)), \\ x(0, t) = x(\pi, t) = 0. \end{cases} \tag{3.25}$$

Define $A := \dfrac{\partial^2}{\partial \zeta^2}$, with its domain

$$\mathcal{D}(A) := \big\{ x \in X : x, x' \text{ are absolutely continuous, } x'' \in X,$$
$$x(0) = x(\pi) = 0 \big\}.$$

Thus, Eq. (3.25) can be abstracted in the form (3.1) with $\zeta(t) := \zeta(t, \cdot)$, $\alpha = -1/2, \beta = 1$ and $\mu = 2$. Since A generates an immediately norm continuous, with $\sigma(A) = \{-n^2 : n \in \mathbb{N}\}$, then solutions to the equation

$$\frac{\lambda(\lambda+1)^2}{(\lambda+1)^2 - \frac{1}{2}} = -n^2,$$

are given by

$$\begin{cases} \lambda_{n,1} = \dfrac{-(n^2+2)}{3} - \dfrac{6a_n}{c_n} + \dfrac{1}{6}c_n, \\[2mm] \lambda_{n,2} = \dfrac{-(n^2+2)}{3} + 3\dfrac{a_n}{c_n} - \dfrac{c_n}{12} + \dfrac{\sqrt{3}}{2}\left(\dfrac{1}{6}c_n + 6\dfrac{a_n}{c_n}\right)i, \\[2mm] \lambda_{n,3} = \dfrac{-(n^2+2)}{3} + 3\dfrac{a_n}{c_n} - \dfrac{c_n}{12} - \dfrac{\sqrt{3}}{2}\left(\dfrac{1}{6}c_n + 6\dfrac{a_n}{c_n}\right)i, \end{cases}$$

for all $n \geq 1$, where

$$\begin{cases} a_n = -\dfrac{1}{9} + \dfrac{2}{9}n^2 - \dfrac{1}{9}n^4, \\[2mm] c_n = \big(8 + 30n^2 + 24n^4 - 8n^6 + 6n\sqrt{24 + 9n^2 + 72n^4 - 24n^6}\big)^{\frac{1}{3}}. \end{cases}$$

An easy computation shows that

$$\sup\left\{ \mathrm{Re}\lambda : \lambda(\lambda+1)^2\left((\lambda+1)^2 - \frac{1}{2}\right)^{-1} \in \sigma(A) \right\} < 0.$$

Thus, from Proposition 3.2 we conclude that there exists a strongly continuous family $\{\mathcal{T}(t)\}_{t\geq 0} \subseteq \mathcal{B}(X)$ such that $\|\mathcal{T}(t)\| \leq Ce^{-\delta t}$ for some $C, \delta > 0$.

Let $g(t, \phi)(s) := \varepsilon\gamma(t)\varrho(\phi(s))$. Assume that $\varrho > 0$, $\gamma(t)$ is a bounded continuous ω-periodic function, i.e., $\gamma(t + \omega) = \gamma(t)$, and ϱ satisfies the

condition $\varrho(e^{ik\omega}x) = e^{ik\omega}\varrho(x), \|\varrho(x) - \varrho(y)\|_X \le L_\varrho \|x - y\|_X, \ L_\varrho > 0.$
Then we have

$$g(t + \omega, \phi)(s) = \epsilon\gamma(t + \omega)\varrho(e^{ik\omega}e^{-ik\omega}\phi(s)) = e^{ik\omega}\epsilon\gamma(t)\varrho(e^{-ik\omega}\phi(s))$$
$$= e^{ik\omega}g(t, e^{-ik\omega}\phi)(s).$$

We also have

$$\|g(t, \phi) - g(t, \psi)\|^2_{L^2[0,\pi]} \le \epsilon^2|\gamma(t)|^2 L_\varrho^2 \|\phi(s) - \psi(s)\|^2_{L^2[0,\pi]}.$$

Then Eq. (3.25) has a unique S-asymptotically Bloch type periodic mild solution on $\mathbb{R}$ by Theorem 3.8 if we choose $\epsilon > 0$ such that $\delta \ge \epsilon\|\gamma\|_\infty L_\varrho \mathcal{C}.$

Chapter 4

Bloch-Type Periodic Solutions to Multi-Term Fractional Evolution Equations

This chapter is mainly concerned with the existence of generalized Bloch-type periodic mild solutions to the following semilinear multi-term fractional evolution equation

$$\partial_t^\alpha u(t) = Au(t) + \partial_t^{\alpha-\beta} f(t, u(t)), \quad t \in \mathbb{R}, \tag{4.1}$$

where A is a closed linear operator defined in a Banach space X, $1 < \alpha$, $\beta < 2$, $f \in BC(\mathbb{R} \times X, X)$, and ∂_t^α denotes the Weyl fractional derivative.

Fractional differential equations describe several physical and biological processes. Some examples include studies in electrochemistry, electromagnetism, viscoelasticity, heredity of materials, rheology, among others. Please refer to Abbas *et al.* (2012), Agrawal *et al.* (2007), Bachir *et al.* (2021), Cui and Sun (2021a,b), Hilfer (2000), Kilbas *et al.* (2006), Liu and Li (2015), Lü and Zuazua (2016), Mainardi (2010), Ponce (2020b, 2020c, 2020d), Wang (2018), Wang *et.al* (2019), Yan and Jia (2015, 2017), Zhou *et al.* (2016), Zhou (2016), and Zhou and He (2021) for more details. We here recall that for a given function $g : \mathbb{R} \to X$, the *Weyl fractional integral* of order $\alpha > 0$ is defined by

$$\partial_t^{-\alpha} g(t) := \frac{1}{\Gamma(\alpha)} \int_{-\infty}^t (t-s)^{\alpha-1} g(s) ds, \quad t \in \mathbb{R},$$

when this integral is convergent. The *Weyl fractional derivatives* $\partial_t^\alpha g$ of order $\alpha > 0$ is defined by

$$\partial_t^\alpha g(t) := \frac{d^n}{dt^n} \partial_t^{-(n-\alpha)} g(t), \quad t \in \mathbb{R},$$

where $n = [\alpha] + 1$, and $[\alpha]$ denotes the integer part of α. It is known that $\partial_t^\alpha \partial_t^{-\alpha} g = g$ for any $\alpha > 0$, and $\partial_t^n = \frac{d^n}{dt^n}$ holds with $n \in \mathbb{N}$.

The existence and qualitative properties of mild solutions to Eq. (4.1) in case $\beta = 1$ have been widely studied in the last years, see for instance Araya and Lizama (2008), Chang *et al.* (2012), Chang and Luo (2015), Chang and Wei (2021b), Cuevas and Lizama (2008), Cuevas and Souza (2009), Ponce (2013b), Wang *et.al* (2012), Wei and Chang (2022), Zhao *et al.* (2013a) and references therein. In these mentioned papers, the operator A is assumed to be a ϖ-sectorial operator of angle θ. In this case, A generates a resolvent family $\{R_\alpha(t)\}_{t\geq 0}$ (see Cuesta, 2007) which satisfies

$$\|R_\alpha(t)\| \leq \frac{C}{1 + |\varpi|t^\alpha}, \quad \text{for all} \quad t \geq 0,$$

where C is a positive constant depending only upon α and θ. The well-posedness, regularity and asymptotic behavior of continuous and discrete solutions to a linear fractional integro-differential equation with varying order in time was also discussed by Cuesta and Ponce (2018).

Ponce (2020b) studied the existence, asymptotic behavior and uniform p-integrability of the (α, β)-resolvent family $\{S_{\alpha,\beta}(t)\}_{t\geq 0}$ generated by sectorial operators in Banach spaces, and further established the asymptotic behavior of solutions to the following fractional Cauchy problems

$$\begin{cases} \partial_c^\alpha u(t) = Au(t) + f(t), & t \geq 0, \\ u(0) = x, \\ u'(0) = y, \end{cases}$$

and

$$\begin{cases} \partial_r^\alpha u(t) = Au(t) + f(t), & t \geq 0, \\ (g_{2-\alpha} * u)(0) = x, \\ (g_{2-\alpha} * u)'(0) = y, \end{cases}$$

where f is a suitable function, A is a closed and linear operator defined in a Banach space X, $x, y \in X$, for $1 < \alpha < 2$, ∂_c^α and ∂_r^α denote, respectively, the Caputo and Riemann-Liouville fractional derivatives (see Appendix A or Kilbas *et al.*, 2006; Zhou *et al.*, 2016), and for $\mu > 0$, $g_\mu(t) := t^{\mu-1}/\Gamma(\mu)$, and the notation $*$ denotes the usual finite convolution. By the asymptotic behavior and uniform 1-integrability of the (α, β)-resolvent family $\{S_{\alpha,\beta}(t)\}_{t\geq 0}$, Chang and Ponce (2021) investigated the existence and uniqueness of bounded mild solutions such as asymptotically periodic, asymptotically almost periodic (automorphic), to Eq. (4.1). In this chapter, we continue to consider the existence and uniqueness of generalized Bloch-type periodic mild solutions to Eq. (4.1) as a supplement.

4.1 Asymptotic Behavior and Uniform Integrability of the Resolvent Family

In this section, we recall some results on the existence, asymptotic behavior and uniform integrability of the resolvent family $\{S_{\alpha,\beta}(t)\}_{t\geq0}$, most of which can be found in Ponce (2020b).

We recall that a closed and densely defined operator A, defined on a Banach space $(X, \|\cdot\|)$, is said to be ϖ-*sectorial of angle* ϕ, if there exist $\phi \in [0, \pi/2)$ and $\varpi \in \mathbb{R}$ such that its resolvent operator $R(\lambda, A) = (\lambda - A)^{-1}$ exists in the sector

$$\varpi + \sum_\phi := \left\{\varpi + \lambda : \lambda \in \mathbb{C}, |\arg(\lambda)| < \frac{\pi}{2} + \phi\right\} \setminus \{\varpi\}$$

and

$$\|R(\lambda, A)\| \leq \frac{M}{|\lambda - \varpi|},$$

for all $\lambda \in \varpi + \Sigma_\phi$. In case $\varpi = 0$ we say that A is sectorial of angle $\phi + \pi/2$. More details on sectorial operators can be found in Haase (2006).

Definition 4.1. Let A be a closed linear operator with the domain $D(A)$, defined on a Banach space X, $1 \leq \alpha \leq 2$ and $0 < \beta \leq 2$. We say that the operator A is the generator of an (α, β)-resolvent family, if there exists $\widetilde{\omega} \geq 0$ and a strongly continuous and exponentially bounded function $S_{\alpha,\beta} : [0, \infty) \to \mathcal{B}(X)$ such that $\{\lambda^\alpha : \mathrm{Re}\lambda > \widetilde{\omega}\} \subset \rho(A)$, and for all $x \in X$,

$$\lambda^{\alpha-\beta}(\lambda^\alpha - A)^{-1}x = \int_0^\infty e^{-\lambda t} S_{\alpha,\beta}(t)x\,dt, \quad \mathrm{Re}\lambda > \widetilde{\omega}.$$

In this case, $\{S_{\alpha,\beta}(t)\}_{t\geq0}$ is called the (α, β)-*resolvent family* generated by the operator A.

Remark 4.1 (Ponce, 2016). From the definition of (a, k)-regularized families in Lizama (2000), it is observed that $t \mapsto S_{\alpha,\beta}(t)$, is a (g_α, g_β)-regularized family. Moreover, $S_{\alpha,\beta}(t)$ satisfies the functional equation (see, Lizama and Poblete, 2012):

$$S_{\alpha,\beta}(s)(g_\alpha * S_{\alpha,\beta})(t) - (g_\alpha * S_{\alpha,\beta})(s)S_{\alpha,\beta}(t)$$

$$= g_\beta(s)(g_\alpha * S_{\alpha,\beta})(t) - g_\beta(t)(g_\alpha * S_{\alpha,\beta})(s),$$

for all $t, s \geq 0$, and, if the operator A with domain $D(A)$ is the infinitesimal generator of an (α, β)-resolvent family, then for all $x \in D(A)$ we have

$$Ax = \lim_{t\to0^+} \frac{S_{\alpha,\beta}(t)x - g_\beta(t)x}{g_{\alpha+\beta}(t)}.$$

For instance, $S_{1,1}(t)$ corresponds to a C_0-semigroup, $S_{2,1}(t)$ to a cosine family and $S_{2,2}(t)$ is a sine family. It is noticed that when $A = \varrho I$, where $\varrho \in \mathbb{C}$ and I denotes the identity operator, then by the uniqueness of the Laplace transform, $S_{\alpha,\beta}(t)$ corresponds to the function $t^{\beta-1}E_{\alpha,\beta}(\varrho t^{\alpha})$. Finally, for $0 < \alpha < 1, \beta \geq \alpha$, we define $\{S_{\alpha,\beta}(t)\}_{t \geq 0}$ by

$$S_{\alpha,\beta}(t)f(s) := \int_0^s f(s-r)\varphi_{\alpha,\beta-\alpha}(t,r)dr,$$

where $s \in \mathbb{R}_+$, $f \in L^1(\mathbb{R}_+)$ and the function $\varphi_{a,b}(t,r)$ is defined by

$$\varphi_{a,b}(t,r) := t^{b-1}W_{-a,b}(-rt^{-a}), \quad a > 0, \ b \geq 0.$$

Here

$$W_{-a,b}(z) := \sum_{n=0}^{\infty} \frac{z^n}{n!\Gamma(-an+b)}, \quad (z \in \mathbb{C})$$

denotes the Wright function. Then, $\{S_{\alpha,\beta}(t)\}_{t \geq 0}$ is an (α, β)-resolvent family on the Banach space $X = L^1(\mathbb{R}_+)$ generated by $A = -\frac{d}{dt}$. See Abadias and Miana (2015, Example 11).

We have also the following result, see Ponce (2020b, Proposition 3.3).

Proposition 4.1. *Let $S_{\alpha,\beta}(t)$ be the (α, β)-resolvent family generated by A for $1 \leq \alpha, \beta \leq 2$. Then:*

(1) $S_{\alpha,\beta}(t)x \in D(A)$ *and* $S_{\alpha,\beta}(t)Ax = AS_{\alpha,\beta}(t)x$ *for all* $x \in D(A)$ *and* $t \geq 0$.

(2) *If* $x \in D(A)$ *and* $t \geq 0$, *then*

$$S_{\alpha,\beta}(t)x = g_\beta(t)x + \int_0^t g_\alpha(t-s)AS_{\alpha,\beta}(s)xds. \tag{4.2}$$

(3) *If* $x \in X, t \geq 0$, *then*

$$\int_0^t g_\alpha(t-s)S_{\alpha,\beta}(s)xds \in D(A),$$

and

$$S_{\alpha,\beta}(t)x = g_\beta(t)x + A\int_0^t g_\alpha(t-s)S_{\alpha,\beta}(s)xds.$$

Particularly, $S_{\alpha,\beta}(0) = g_\beta(0)I$.

The next generation result (analogous to the Hille–Yosida Theorem for C_0-semigroups) is contained in Lizama (2000, Theorem 3.4). See Ponce (2020b, Theorem 3.4).

Theorem 4.1. *Let A be a closed linear densely defined operator in a Banach space X. Suppose that $1 < \alpha < 2$ and $\beta \geq 1$ such that $\alpha - \beta + 1 > 0$. Then the following assertions are equivalent:*

(1) *The operator A generates an (α, β)-resolvent family $\{S_{\alpha,\beta}(t)\}_{t \geq 0}$ which satisfies $\|S_{\alpha,\beta}(t)\| \leq M e^{\mu t}$ for all $t \geq 0$ and for some constants $M > 0$ and $\mu \in \mathbb{R}$.*
(2) *There exist constants $\mu \in \mathbb{R}$ and $M > 0$ such that $\lambda^\alpha \in \rho(A)$ for all λ with $\mathrm{Re}\lambda > \mu$ and $H(\lambda) := \lambda^{\alpha-\beta} (\lambda^\alpha - A)^{-1}$ satisfies*

$$\|H^{(n)}(\lambda)\| \leq \frac{Mn!}{(\lambda - \mu)^{n+1}},$$

for all $\mathrm{Re}\lambda > \mu$ and $n \in \mathbb{N} \cup \{0\}$.

The next result gives sufficient conditions on α, β and A to obtain generators of an (α, β)-resolvent family $\{S_{\alpha,\beta}(t)\}_{t \geq 0}$. See Ponce (2020b, Theorem 3.5).

Theorem 4.2. *Let $1 < \alpha < 2$ and $\beta \geq 1$ such that $\alpha - \beta + 1 > 0$. Assume that A is ϖ-sectorial of angle $\frac{(\alpha-1)\pi}{2}$, where $\varpi < 0$. Then A generates an exponentially bounded (α, β)-resolvent family $\{S_{\alpha,\beta}(t)\}_{t \geq 0}$.*

Proof. It will be shown that $\lambda^\alpha \in \rho(A)$ for all λ with $\mathrm{Re}\lambda > 0$, and there exists a constant $C > 0$ such that the function $H(\lambda) := \lambda^{\alpha-\beta}(\lambda^\alpha - A)^{-1}$ satisfies the estimate $\|\lambda H(\lambda)\| + \|\lambda^2 H'(\lambda)\| \leq C$, for all $\mathrm{Re}\lambda > 0$. Now for $\lambda = re^{i\theta}$ with $|\theta| < \frac{\pi}{2}$ and $r > 0$, we define $h(\lambda) := \lambda^\alpha$. We observe that

$$\arg(h(re^{i\theta})) = \mathrm{Im}(\ln(h(re^{i\theta}))) = \mathrm{Im} \int_0^\theta \frac{d}{dt} \ln(h(re^{it})) dt$$

$$= \mathrm{Im} \int_0^\theta \frac{h'(re^{it}) i re^{it}}{h(re^{it})} dt.$$

Since $\lambda \frac{h'(\lambda)}{h(\lambda)} = \alpha$, we obtain

$$\left| \mathrm{Im} \int_0^\theta \frac{h'(re^{it}) i re^{it}}{h(re^{it})} dt \right| \leq \int_0^\theta \left| \frac{h'(re^{it}) i re^{it}}{h(re^{it})} \right| dt \leq \alpha \theta$$

$$\leq \frac{(\alpha - 1)\pi}{2} + \frac{\pi}{2}.$$

Thus, $h(\lambda) \in \Sigma_{\frac{(\alpha-1)\pi}{2}}$ for all $\text{Re}\lambda > 0$, and H is well defined. Since A is a ϖ-sectorial operator, there exists $M > 0$ such that

$$\|\lambda H(\lambda)\| \le M \frac{|\lambda|^{\alpha-\beta+1}}{|\lambda^\alpha - \varpi|}, \quad \text{for all } \text{Re}\lambda > 0.$$

Since $\beta \ge 1$ and $\alpha - \beta + 1 > 0$, we obtain

$$\|\lambda H(\lambda)\| \le M, \quad \text{for all } \text{Re}\lambda > 0.$$

A simple computation gives

$$\lambda^2 H'(\lambda) = (\alpha - \beta)\lambda H(\lambda) + \alpha^2 \lambda H(\lambda)\lambda^\alpha(\lambda^\alpha - A)^{-1},$$

and thus

$$\|\lambda^2 H'(\lambda)\| \le |\alpha - \beta|\|\lambda H(\lambda)\| + \alpha^2\|\lambda H(\lambda)\lambda^\alpha(\lambda^\alpha - A)^{-1}\|$$

$$\le |\alpha - \beta|M + \frac{\alpha^2 M^2 |\lambda|^\alpha}{|\lambda^\alpha - \varpi|} \le (|\alpha - \beta| + \alpha^2 M)M.$$

Therefore,

$$\|\lambda H(\lambda)\| + \|\lambda^2 H'(\lambda)\| \le M + |\alpha - \beta|M + \alpha^2 M^2, \quad \text{for all } \text{Re}\lambda > 0.$$

We conclude by Lemma 1.10 (cf. Prüss, 1993, Proposition 0.1) and Theorem 4.1 that the operator A generates an exponentially bounded (α, β)-resolvent family. $\square$

The following result is involved in asymptotic behavior of the resolvent family $\{S_{\alpha,\beta}(t)\}_{t\ge 0}$. See Ponce (2020b, Theorem 3.6).

Theorem 4.3. *Let $1 < \alpha < 2$ and $\beta \ge 1$ such that $\alpha - \beta + 1 > 0$. Assume that A is ϖ-sectorial of angle $\frac{(\alpha-1)}{2}\pi$, where $\varpi < 0$. Then there exists a constant $C > 0$, depending only on α and β, such that*

$$\|S_{\alpha,\beta}(t)\| \le \frac{Ct^{\beta-1}}{1 + |\varpi|t^\alpha}, \quad \text{for all } t > 0. \tag{4.3}$$

Proof. Since A is ϖ-sectorial of angle $\theta := \frac{(\alpha-1)}{2}\pi$, it follows from Theorem 4.2 that $h(\lambda) := \lambda^\alpha \in \rho(A)$ for all $\text{Re}\lambda > 0$, and $\|(\lambda^\alpha - A)^{-1}\| \le \frac{M}{|\lambda^\alpha - \varpi|}$, for all $\lambda \in \mathbb{C}, \text{Re}\lambda > 0$. Next, we write

$$S_{\alpha,\beta}(t) = \frac{1}{2\pi i} \int_\gamma e^{\lambda t} \lambda^{\alpha-\beta}(\lambda^\alpha - A)^{-1}d\lambda, \tag{4.4}$$

where γ is a positively oriented path lying inside the sector $\varpi + \Sigma_\theta$, whose support Γ is the set of $\lambda \in \mathbb{C}$ such that λ^α belongs to the boundary of B_δ, where $B_\delta := \{\delta + |\varpi| + \Sigma_\theta\} + \{\delta + \Sigma_\phi\}$, with $\delta > 0$ and $0 < \phi < \theta$.

With this definition of γ, $(\lambda^\alpha - A)^{-1}$ is well defined and the representation (4.4) of $S_{\alpha,\beta}(t)$ makes sense. We split γ into two parts γ_1, γ_2, whose supports Γ_1 and Γ_2 are the sets formed by the complex numbers λ such that λ^α lies on the intersection of Γ and the boundaries of $|\varpi| + 1/t^\alpha + \Sigma_\theta$ and $1/t^\alpha + \Sigma_\phi$ respectively, i.e.,

$$\Gamma_1 = \Gamma \cap \overline{\left\{ |\varpi| + \frac{1}{t^\alpha} + \Sigma_\theta \right\}} \quad \text{and} \quad \Gamma_2 = \Gamma \cap \overline{\left\{ \frac{1}{t^\alpha} + \Sigma_\phi \right\}}.$$

Therefore, $\Gamma = \Gamma_1 \cup \Gamma_2$ and $S_{\alpha,\beta}(t) = I_1(t) + I_2(t)$, for $t \geq 0$, where

$$I_j(t) := \frac{1}{2\pi i} \int_{\gamma_j} e^{\lambda t} \lambda^{\alpha - \beta} (\lambda^\alpha - A)^{-1} d\lambda, \quad \text{for } j = 1, 2.$$

We now estimate the integrals $I_1(t)$ and $I_2(t)$. For the integral $I_1(t)$ we have

$$\|I_1(t)\| \leq \frac{1}{2\pi} \int_{\gamma_1} |e^{\lambda t}| |\lambda|^{\alpha - \beta} \|(\lambda^\alpha - A)^{-1}\| |d\lambda|$$

$$\leq \frac{M}{2\pi} \int_{\gamma_1} |e^{\lambda t}| \frac{|\lambda|^{\alpha - \beta}}{|\lambda^\alpha - \varpi|} |d\lambda|.$$

We define $\lambda_{\min}$ as the complex $\lambda \in \mathbb{C}$ such that $\text{Im}(\lambda) > 0$, and $|\lambda_{\min}^\alpha - \varpi| = \text{dist}(L, \varpi)$, where L is the line passing by $(|\varpi| + 1/t^\alpha, 0)$ and the intersection of Γ_1 and Γ_2. For $\lambda \in \Gamma_1$ we have that

$$\begin{cases} |\lambda_{\min}^\alpha - \varpi| \leq |\lambda^\alpha - \varpi|; \\ \cos(\theta) = \sin(\frac{\pi}{2} - \theta) = \frac{|\lambda_{\min}^\alpha - \varpi|}{|\varpi| + \frac{1}{t^\alpha}} \leq \frac{|\lambda^\alpha - \varpi|}{|\varpi| + \frac{1}{t^\alpha}}. \end{cases}$$

Therefore, if $\lambda \in \Gamma_1$, then

$$\frac{1}{|\lambda^\alpha - \varpi|} \leq \frac{t^\alpha}{\cos(\theta)(1 + |\varpi| t^\alpha)}.$$

Hence,

$$\|I_1(t)\| \leq \frac{M t^\alpha}{2\pi \cos(\theta)(1 + |\varpi| t^\alpha)} \int_{\gamma_1} |e^{\lambda t}| |\lambda|^{\alpha - \beta} |d\lambda|$$

$$\leq \frac{M t^\alpha}{\pi \cos(\theta)(1 + |\varpi| t^\alpha)} \int_0^\infty e^{-t \cos(\theta) s} s^{\alpha - \beta} ds$$

$$= \frac{C_\theta t^{\beta - 1}}{1 + |\varpi| t^\alpha},$$

where $C_\theta = \frac{M\Gamma(\alpha-\beta+1)}{\pi(\cos(\theta))^{\alpha-\beta+2}}$. Similarly, if $\lambda \in \Gamma_2$, then

$$\frac{1}{|\lambda^\alpha - \varpi|} \leq \frac{t^\alpha}{\cos(\phi)(1 + |\varpi|t^\alpha)},$$

and thus

$$\|I_2(t)\| \leq \frac{Mt^\alpha}{2\pi \cos(\phi)(1 + |\varpi|t^\alpha)} \cdot \int_{\gamma_1} |e^{\lambda t}||\lambda|^{\alpha-\beta}|d\lambda|$$

$$\leq \frac{Mt^\alpha}{\pi \cos(\phi)(1 + |\varpi|t^\alpha)} \int_0^\infty e^{-t\cos(\phi)s} s^{\alpha-\beta} ds$$

$$= \frac{C_\phi t^{\beta-1}}{1 + |\varpi|t^\alpha},$$

where $C_\phi := \frac{M\Gamma(\alpha-\beta+1)}{\pi(\cos(\phi))^{\alpha-\beta+2}}$. Thus, there exists a constant $C > 0$, depending only on α, β, such that

$$\|S_{\alpha,\beta}(t)\| \leq \frac{Ct^{\beta-1}}{1 + |\varpi|t^\alpha}$$

for all $t \geq 0$. $\qquad\qquad\qquad\qquad\qquad\qquad\qquad\qquad\qquad\qquad\qquad$ □

Definition 4.2. The family $\{S(t)\}_{t\geq 0} \subset \mathcal{B}(X)$ is called *asymptotically stable* if $\|S(t)\| \to 0$ as $t \to \infty$.

We have the following corollary via Theorem 4.3. See Ponce (2020b, Corollaries 3.8 and 3.9).

Corollary 4.1. *Let* $(X, \|\cdot\|)$ *be a Banach space. Then the following assertions are satisfied.*

(a) *If* $1 < \alpha < 2$ *and* $\beta \geq 1$ *are such that* $\alpha - \beta + 1 > 0$, *and* A *is* ϖ-*sectorial operator of angle* $\theta = \frac{(\alpha-1)}{2}\pi$, *where* $\varpi < 0$, *then* $\{S_{\alpha,\beta}(t)\}_{t\geq 0}$ *is asymptotically stable.*

(b) *If* $1 \leq \beta < \alpha < 2$ *and* A *is* ϖ-*sectorial of angle* $\theta = \frac{(\alpha-1)}{2}\pi$, *where* $\varpi < 0$, *then* $\{S_{\alpha,\beta}(t)\}_{t\geq 0}$ *is uniformly 1-integrable.*

Proof. The assertion (a) can be obtained from (4.3) in Theorem 4.3. For assertion (b), the condition $1 \leq \beta < \alpha < 2$ implies $\alpha - \beta + 1 > 0$ and $1 - \frac{\beta}{\alpha} > 0$. Thus it follows from (4.3) that

$$\int_0^\infty \|S_{\alpha,\beta}(t)\| dt \leq \int_0^\infty \frac{Ct^{\beta-1}}{1 + |\varpi|t^\alpha} dt = \frac{C}{\alpha}|\varpi|^{-\beta/\alpha}\mathbf{B}\left(\frac{\beta}{\alpha}, 1 - \frac{\beta}{\alpha}\right) < \infty,$$

where $\mathbf{B}(\cdot, \cdot)$ denotes the Beta function. $\qquad\qquad\qquad\qquad\qquad\qquad\qquad$ □

4.2   Linear Fractional Evolution Equation

In this section, we consider the following linear equation corresponding to Eq. (4.1). Some of results can be found in [Chang and Ponce (2021)].

$$\partial_t^\alpha u(t) = Au(t) + \partial_t^{\alpha-\beta} f(t), \quad t \in \mathbb{R}. \tag{4.5}$$

We give the following definition.

Definition 4.3. A function $u \in C(\mathbb{R}, X)$ is called a *mild solution* to Eq. (4.5) if the function $s \mapsto S_{\alpha,\beta}(t-s)f(s)$ is integrable on $(-\infty, t)$ for each $t \in \mathbb{R}$ and

$$u(t) = \int_{-\infty}^t S_{\alpha,\beta}(t-s)f(s)ds, \quad t \in \mathbb{R}.$$

It is noted that Eq. (4.5) can be regarded as the limiting equation of the following integrodifferential equation with singular kernels

$$\begin{cases} v'(t) = \int_0^t \dfrac{(t-s)^{\alpha-2}}{\Gamma(\alpha-1)} Av(s)ds + \int_0^t \dfrac{(t-s)^{\beta-2}}{\Gamma(\beta-1)} f(s)ds, \quad t \ge 0 \\ v(0) = v_0, \ v_0 \in X, \end{cases} \tag{4.6}$$

in the sense that the mild solution to Eq. (4.6) converges to the mild solution of Eq. (4.5) as $t \to \infty$. In fact, if $\varpi < 0$ and A is a ϖ-sectorial operator of angle $\theta = \frac{(\alpha-1)}{2}\pi$, then taking Laplace transform in Eq. (4.6) we obtain

$$\lambda \hat{v}(\lambda) - v(0) = \frac{1}{\lambda^{\alpha-1}} A\hat{v}(\lambda) + \frac{1}{\lambda^{\beta-1}} \hat{f}(\lambda), \quad \mathrm{Re}\lambda > 0,$$

which is equivalent to

$$(\lambda^\alpha - A)\hat{v}(\lambda) = \lambda^{\alpha-1} v(0) + \lambda^{\alpha-\beta} \hat{f}(\lambda), \quad \mathrm{Re}\lambda > 0.$$

Hence the solution of Eq. (4.6) can be written as

$$v(t) = S_{\alpha,1}(t)v_0 + \int_0^t S_{\alpha,\beta}(t-s)f(s)ds, \quad t \ge 0, \tag{4.7}$$

where $\{S_{\alpha,\beta}(t)\}_{t \ge 0}$ is the family of operators given by

$$S_{\alpha,\beta}(t) := (g_{\beta-1} * S_{\alpha,1})(t).$$

On the other hand, by Corollary 4.1(b) (see also Ponce, 2020b, Corollary 3.9) the function $t \mapsto S_{\alpha,\beta}(t)$ is uniformly 1-integrable and thus if $f \in BC(\mathbb{R}, X)$, then the mild solution to Eq. (4.5) is given by

$$u(t) = \int_{-\infty}^t S_{\alpha,\beta}(t-s)f(s)ds.$$

Since

$$v(t) - u(t) = S_{\alpha,1}(t)v_0 - \int_t^\infty S_{\alpha,\beta}(s)f(t-s)ds,$$

we conclude by Corollary 4.1(a) (see also Ponce, 2020b, Corollary 3.8), that $v(t) - u(t) \to 0$ as $t \to \infty$.

Let $1 < \alpha < 2$, $\beta \geq 1$ be such that $\alpha - \beta + 1 > 0$, $\varpi < 0$ and assume that A is a ϖ-sectorial operator of angle $\theta = \frac{(\alpha-1)}{2}\pi$. By Theorem 4.2, the operator A generates a resolvent family $\{S_{\alpha,\beta}(t)\}_{t\geq0}$. Assume that $f \in BC(\mathbb{R}, X)$ and define the function $\phi(t)$ by

$$\phi(t) := \int_{-\infty}^t S_{\alpha,\beta}(t-s)f(s)ds, \quad t \in \mathbb{R}. \tag{4.8}$$

By Corollary 4.1(b) we have $||\phi||_\infty \leq ||S_{\alpha,\beta}||_1 \, ||f||_\infty$. If $f(t) \in D(A)$ for all $t \in \mathbb{R}$, then $\phi(t) \in D(A)$ for all $t \in \mathbb{R}$ (see Lemma 1.4 or Arendt *et al.*, 2001, Proposition 1.1.7). Assume that $\partial_t^\alpha \phi$ exists. It follows by Proposition 4.1 and Fubini's theorem that

$$\partial_t^\alpha \phi(t) = \frac{d^n}{dt^n} \int_{-\infty}^t g_{n-\alpha}(t-s)\phi(s)ds$$

$$= \frac{d^n}{dt^n} \int_{-\infty}^t g_{n-\alpha}(t-s) \int_{-\infty}^s S_{\alpha,\beta}(s-r)f(r)drds$$

$$= \frac{d^n}{dt^n} \int_{-\infty}^t g_{n-\alpha}(t-s)$$

$$\times \int_{-\infty}^s \Big[g_\beta(s-r)f(r) + (g_\alpha * AS_{\alpha,\beta})(s-r)f(r)\Big]drds$$

$$= \frac{d^n}{dt^n} \int_{-\infty}^t g_{n-\alpha}(t-s)\partial_t^{-\beta}f(s)ds + \frac{d^n}{dt^n}\int_{-\infty}^t g_{n-\alpha}(t-s)$$

$$\times \int_{-\infty}^s \int_0^{s-r} g_\alpha(s-r-v)AS_{\alpha,\beta}(v)f(r)dvdrds$$

$$= \partial_t^{\alpha-\beta}f(t) + \frac{d^n}{dt^n}\int_{-\infty}^t g_{n-\alpha}(t-s)$$

$$\times \int_{-\infty}^{s} \int_{r}^{s} g_\alpha(s-w) A S_{\alpha,\beta}(w-r) f(r) dw dr ds$$

$$= \partial_t^{\alpha-\beta} f(t) + \frac{d^n}{dt^n} \int_{-\infty}^{t} g_{n-\alpha}(t-s)$$

$$\times \int_{-\infty}^{s} \int_{-\infty}^{w} g_\alpha(s-w) A S_{\alpha,\beta}(w-r) f(r) dr dw ds$$

$$= \partial_t^{\alpha-\beta} f(t) + \frac{d^n}{dt^n} \int_{-\infty}^{t} g_{n-\alpha}(t-s)$$

$$\times \int_{-\infty}^{s} g_\alpha(s-w) A \phi(w) dw ds = \partial_t^{\alpha-\beta} f(t) + A\phi(t),$$

for all $t \in \mathbb{R}$. This means that, ϕ is a (strong) solution to Eq. (4.5). We recall that a function $u \in C(\mathbb{R}, X)$ is called a strong solution of Eq. (4.5) on $\mathbb{R}$ if $u \in C(\mathbb{R}, D(A))$, the fractional derivative of u, $\partial_t^\alpha u$, exists and Eq. (4.5) holds for all $t \in \mathbb{R}$. If merely $u(t)$ belongs to X instead of the $D(A)$, then u is a mild solution to Eq. (4.5) according to Definition 4.3. As a consequence of the above computation we have the following result.

Theorem 4.4. *Let $1 \leq \beta < \alpha < 2$ and $\varpi < 0$. Assume further that A is a ϖ-sectorial operator of angle $\theta = \frac{(\alpha-1)}{2}\pi$. Then for each $f \in \mathcal{N}(X)$ there exists a unique mild solution $u \in \mathcal{N}(X)$ to Eq. (4.5) which is given by*

$$u(t) = \int_{-\infty}^{t} S_{\alpha,\beta}(t-s) f(s) ds, \quad t \in \mathbb{R}.$$

Proof. By Theorem 4.2, the operator A generates a resolvent family $\{S_{\alpha,\beta}(t)\}_{t \geq 0}$ and by Corollary 4.1(b) (see also Ponce, 2020b, Corollary 3.9), the function $t \mapsto S_{\alpha,\beta}(t)$ is uniformly 1-integrable. By Theorems 2.1, 2.2, 2.7, 2.10, 2.13 of Chapter 2, the function defined by $u(t) = \int_{-\infty}^{t} S_{\alpha,\beta}(t-s) f(s) ds$ belongs to $\mathcal{N}(X)$ and it is the mild solution to Eq. (4.5). $\square$

4.3   Semilinear Fractional Evolution Equation

In this section, we investigate the existence of mild solutions in $\mathcal{N}(X)$ for the semilinear Eq. (4.1). In what follows, we assume that the following conditions hold:

(H_A) Let the operator A be a ϖ-sectorial operator of angle $\theta = \frac{(\alpha-1)}{2}\pi$ with $\varpi < 0$, and $1 \leq \beta < \alpha < 2$.

(Hf_1) Let $f \in BC(\mathbb{R} \times X, X)$ and there exists a constant $L > 0$ such that for all $t \in \mathbb{R}$ and $u, v \in X$,

$$\|f(t, u) - f(t, v)\| \leq L\|u - v\|.$$

(Hf_2) Let $f \in BC(\mathbb{R} \times X, X)$ and there exists a function $\mathfrak{L}(\cdot) \in L^1(\mathbb{R}, \mathbb{R}_+)$ such that for all $t \in \mathbb{R}$ and $u, v \in X$,

$$\|f(t, u) - f(t, v)\| \leq \mathfrak{L}(t)\|u - v\|.$$

Remark 4.2. By Theorem 4.2 and Corollary 4.1(b), it is concluded that the condition (H_A) ensures the existence of a resolvent family $\{S_{\alpha,\beta}(t)\}_{t\geq 0}$, which is uniformly 1-integrable and satisfies

$$\int_0^\infty \|S_{\alpha,\beta}(t)\|dt \leq \frac{C}{\alpha}|\varpi|^{-\beta/\alpha}\mathbf{B}\left(\frac{\beta}{\alpha}, 1 - \frac{\beta}{\alpha}\right) < \infty. \tag{4.9}$$

Definition 4.4. A function $u \in C(\mathbb{R}, X)$ is called a *mild solution* to Eq. (4.1) if the function $s \mapsto S_{\alpha,\beta}(t - s)f(s, u(s))$ is integrable on $(-\infty, t)$ for each $t \in \mathbb{R}$ and

$$u(t) = \int_{-\infty}^t S_{\alpha,\beta}(t - s)f(s, u(s))ds, \quad t \in \mathbb{R}.$$

Theorem 4.5. *Let conditions* (H_A) *and* (Hf_1) *hold. Assume further that* $f = g + h \in BC(\mathbb{R} \times X, X)$ *with* g *satisfying the condition* ($A1$) *in Section 2.3 and* $h \in \mathscr{E}(\mathbb{R} \times X, X)$. *Then Eq.* (4.1) *admits a unique mild solution* $u \in PBP_{\omega,k}(\mathbb{R}, X)$ *if*

$$0 < L < \frac{\alpha}{C}|\varpi|^{\beta/\alpha}\mathbf{B}\left(\frac{\beta}{\alpha}, 1 - \frac{\beta}{\alpha}\right)^{-1},$$

where C *is the constant given in Theorem 4.3, and* $\mathbf{B}(\cdot, \cdot)$ *denotes the Beta function.*

Proof. It follows by Remark 4.2 that A generates a resolvent operator family $\{S_{\alpha,\beta}(t)\}_{t\geq 0}$, which is uniformly 1-integrable and satisfies (4.9). Define the operator $F : PBP_{\omega,k}(\mathbb{R}, X) \to PBP_{\omega,k}(\mathbb{R}, X)$ by

$$(F\phi)(t) := \int_{-\infty}^{t} S_{\alpha,\beta}(t-s)f(s,\phi(s))ds, \quad t \in \mathbb{R}. \tag{4.10}$$

For each $\phi \in PBP_{\omega,k}(\mathbb{R}, X)$, by (1) in Corollary 2.2, the function $s \mapsto f(s,\phi(s))$ belongs to $PBP_{\omega,k}(\mathbb{R}, X)$. It follows from Theorem 4.4 that $F\phi \in PBP_{\omega,k}(\mathbb{R}, X)$ and F is well-defined. For $\phi_1, \phi_2 \in PBP_{\omega,k}(\mathbb{R}, X)$ and $t \in \mathbb{R}$, by (4.9), we have

$$\|(F\phi_1)(t) - (F\phi_2)(t)\|$$
$$\leq \int_{-\infty}^{t} \|S_{\alpha,\beta}(t-s)[f(s,\phi_1(s)) - f(s,\phi_2(s))]\|ds$$
$$\leq \int_{-\infty}^{t} L\|S_{\alpha,\beta}(t-s)\| \cdot \|\phi_1(s) - \phi_2(s)\|ds$$
$$\leq L\|\phi_1 - \phi_2\|_\infty \int_0^\infty \|S_{\alpha,\beta}(r)\|dr$$
$$\leq \frac{LC}{\alpha}|\varpi|^{-\beta/\alpha}\mathbf{B}\left(\frac{\beta}{\alpha}, 1-\frac{\beta}{\alpha}\right)\|\phi_1 - \phi_2\|_\infty,$$

i.e.,

$$\|F\phi_1 - F\phi_2\|_\infty \leq \frac{LC}{\alpha}|\varpi|^{-\beta/\alpha}\mathbf{B}\left(\frac{\beta}{\alpha}, 1-\frac{\beta}{\alpha}\right)\|\phi_1 - \phi_2\|_\infty.$$

This shows that F is a contraction, and thus by the Banach fixed point theorem there exists a unique $u \in PBP_{\omega,k}(\mathbb{R}, X)$ such that $Fu = u$. $\square$

Theorem 4.6. *Let conditions (H_A) and (Hf_2) hold. Suppose further that $f = g+h \in BC(\mathbb{R}\times X, X)$ with g satisfying the condition $(A1)$ in Section 2.3 and $h \in \mathscr{E}(\mathbb{R} \times X, X)$. Then Eq. (4.1) admits a unique mild solution $u \in PBP_{\omega,k}(\mathbb{R}, X)$.*

Proof. It is noticed by Theorem 4.3 that if $t \geq 1$, then

$$\|S_{\alpha,\beta}(t)\| \leq \frac{C}{|\varpi|}\frac{1}{t^{\alpha-\beta+1}} \leq \frac{C}{|\varpi|},$$

and if $0 \le t \le 1$, then

$$\|S_{\alpha,\beta}(t)\| \le \frac{C}{1 + |\varpi|t^\alpha} \le C.$$

Therefore, $\|S_{\alpha,\beta}(t)\| \le \widetilde{C}$, where $\widetilde{C} = \max\left\{C, \frac{C}{|\varpi|}\right\}$.

Define the operator F as (4.10). For each $\phi \in PBP_{\omega,k}(\mathbb{R}, X)$, it follows by Corollary 2.3 of Chapter 2 and Theorem 4.4 that $F\phi \in PBP_{\omega,k}(\mathbb{R}, X)$. So F is well defined. For $u, v \in PBP_{\omega,k}(\mathbb{R}, X)$ and $t \in \mathbb{R}$, we have

$$\|(Fu)(t) - (Fv)(t)\|$$

$$\le \int_{-\infty}^t \|S_{\alpha,\beta}(t - s)[f(s, u(s)) - f(s, v(s))]\| ds$$

$$\le \widetilde{C}\|u - v\|_\infty \int_0^\infty \mathfrak{L}(t - \xi)d\xi$$

$$= \widetilde{C}\|u - v\|_\infty \int_{-\infty}^t \mathfrak{L}(s)ds.$$

Generally, we have

$$\|(F^n u)(t) - (F^n v)(t)\|$$

$$\le \|u - v\|_\infty \frac{(\widetilde{C})^n}{(n - 1)!} \left(\int_{-\infty}^t \mathfrak{L}(s) \left(\int_{-\infty}^s \mathfrak{L}(\xi)d\xi\right)^{n-1} ds\right)$$

$$\le \|u - v\|_\infty \frac{(\widetilde{C})^n}{n!} \left(\int_{-\infty}^t \mathfrak{L}(s)ds\right)^n$$

$$\le \|u - v\|_\infty \frac{(\|\mathfrak{L}\|_1 \widetilde{C})^n}{n!},$$

i.e.,

$$\|F^n u - F^n v\|_\infty \le \frac{(\|\mathfrak{L}\|_1 \widetilde{C})^n}{n!} \|u - v\|_\infty.$$

Since $\frac{(\|\mathfrak{L}\|_1 \widetilde{C})^n}{n!} < 1$ for sufficiently large n, by the Banach contraction principle (see Corollary 1.1 of Chapter 1), the operator F admits a unique fixed point $u \in PBP_{\omega,k}(\mathbb{R}, X)$. $\qquad\square$

Remark 4.3. Let $\rho \in \mathcal{U}_{\text{inv}}$ and take into account Theorem 4.4, Corollaries 2.2(2) and 2.4 of Chapter 2, we can obtain Eq. (4.1) admits a unique mild solution $u \in WPBP_{\omega,k}(\mathbb{R}, X, \rho)$ such as Theorems 4.5–4.6.

Theorem 4.7. *Let conditions (H_A) and (Hf_1) hold. Assume further that $f \in BC(\mathbb{R} \times X, X)$ satisfies the condition $(T1)$ in Theorem 2.5 of Chapter 2 with*

$$0 < L < \frac{\alpha}{C} |\varpi|^{\beta/\alpha} \mathbf{B}\left(\frac{\beta}{\alpha}, 1 - \frac{\beta}{\alpha}\right)^{-1}.$$

Then Eq. (4.1) has a unique mild solution $u \in SABP_{\omega,k}(\mathbb{R}, X)$.

Proof. Define the operator $F : SABP_{\omega,k}(\mathbb{R}, X) \to SABP_{\omega,k}(\mathbb{R}, X)$ by

$$(Fu)(t) := \int_{-\infty}^{t} S_{\alpha,\beta}(t - s)(t - s)f(s, u(s))ds, \ t \in \mathbb{R}.$$

According to Theorems 2.5 of Chapter 2 and 4.4, Fu belongs to $SABP_{\omega,k}(\mathbb{R}, X)$ whenever $u \in SABP_{\omega,k}(\mathbb{R}, X)$. Thus, the operator F is well defined. It remains to show that F is a contraction on $SABP_{\omega,k}(\mathbb{R}, X)$, which can be conducted just as Theorem 4.5. □

Remark 4.4. Based upon Theorems 4.4, 2.8, and 2.11 of Chapter 2, we can obtain that Eq. (4.1) admits a unique mild solution $u \in PSABP_{\omega,k}(\mathbb{R}, X)$ (or $WPSABP_{\omega,k}(\mathbb{R}, X, \rho)$) such as Theorem 4.7, respectively.

Next we investigate the existence of pseudo (ω, k)-Bloch periodic mild solutions to Eq. (4.1) when f is not necessarily Lipschitz continuous with its second variable. We shall use the space defined by Eq. (1.5) in Chapter 1 and the following condition.

(H_C) Let A generate an exponentially bounded resolvent operator family $\{S_{\alpha,\beta}(t)\}_{t\geq 0}$, i.e., $\|S_{\alpha,\beta}(t)\| \leq Me^{\mu t}$ for all $t \geq 0$ and for some constants $M > 0$ and $\mu \in \mathbb{R}$. Assume further that $(\lambda^\alpha - A)^{-1}$ is compact for all $\lambda \in \rho(A)$ with $\lambda > \mu^{\frac{1}{\alpha}}$.

Remark 4.5. The condition (H_A) can ensure the existence of an exponentially bounded resolvent operator family $\{S_{\alpha,\beta}(t)\}_{t\geq 0}$ (see Theorem 4.2). And the condition (H_C) can guarantee that the resolvent operator $S_{\alpha,\beta}(t)$ is compact for all $t > 0$, see Ponce (2016, Theorem 14) (or Theorem A.3 in Appendix A). Define

$$\Phi(t) := \frac{Ct^{\beta-1}}{1 + |\varpi|t^\alpha}, \quad t \geq 0.$$

It is known from (H_A) and Corollary 4.1(b) that $\Phi(t) \in L^1(\mathbb{R}_+)$.

Theorem 4.8. *Let conditions* (H_A) *and* (H_C) *hold. Assume that* $f = g + h \in BC(\mathbb{R} \times X, X)$ *with* g *satisfying the condition* $(A1)$ *in Section 2.3 of Chapter 2,* $h \in \mathscr{E}(\mathbb{R} \times X, X)$. *Suppose further that:*

(Cf1) $f \in BC(\mathbb{R} \times X, X)$ *and* $f(t, x)$ *is uniformly continuous in any bounded subset* $\mathbb{K} \subseteq X$ *uniformly for* $t \in \mathbb{R}$.

(Cf2) *There exists a continuous nondecreasing function* $W_f : \mathbb{R}_+ \to \mathbb{R}_+$ *such that*

$$\|f(t, x)\| \le W_f(\|x\|) \text{ for all } t \in \mathbb{R} \text{ and } x \in X.$$

(Cf3) *For each* $r \ge 0$, *the function* $t \to \int_{-\infty}^{t} \Phi(t - s) W_f(r\mathrm{h}(s)) ds$ *belongs to* $BC(\mathbb{R})$. *We set*

$$\beta(r) = \left\| \int_{-\infty}^{t} \Phi(t - s) W_f(r\mathrm{h}(s)) ds \right\|_{\mathrm{h}}.$$

(Cf4) *For each* $\varepsilon > 0$ *there is* $\delta > 0$ *such that for every* $u, v \in C_{\mathrm{h}}(X)$, $\|u - v\|_{\mathrm{h}} \le \delta$ *implies that*

$$\int_{-\infty}^{t} \Phi(t - s) \| f(s, u(s)) - f(s, v(s)) \| ds \le \varepsilon$$

for all $t \in \mathbb{R}$.

(Cf5) $\liminf_{\xi \to \infty} \frac{\xi}{\beta(\xi)} > 1$.

Then Eq. (4.1) has at least one mild solution $u \in PBP_{\omega,k}(\mathbb{R}, X)$.

Proof. Define the operator $\Lambda : C_{\mathrm{h}}(X) \to C_{\mathrm{h}}(X)$ by

$$(\Lambda u)(t) := \int_{-\infty}^{t} S_{\alpha,\beta}(t - s) f(s, u(s)) ds, \quad t \in \mathbb{R}.$$

It will be shown that Λ has a fixed point in $PBP_{\omega,k}(\mathbb{R}, X)$. For the sake of convenience, we divide the proof into several steps.

(I) For each $u \in C_{\mathrm{h}}(X)$, it follows from the condition (Cf2) that

$$\|(\Lambda u)(t)\| \le \int_{-\infty}^{t} \Phi(t - s) W_f(\|u(s)\|) ds$$

$$\le \int_{-\infty}^{t} \Phi(t - s) W_f(\|u\|_{\mathrm{h}} \mathrm{h}(s)) ds.$$

Thus, the condition (Cf3) implies that $\Lambda : C_{\mathrm{h}}(X) \to C_{\mathrm{h}}(X)$.

(II) The operator Λ is continuous. In fact, for any $\varepsilon > 0$, we take $\delta > 0$ involved in the condition (Cf4). If $u, v \in C_{\hbar}(X)$ with $\|u - v\|_{\hbar} \leq \delta$, then

$$\|(\Lambda u)(t) - (\Lambda v)(t)\| \leq \int_{-\infty}^{t} \Phi(t - s)\|f(s, u(s)) - f(s, v(s))\| ds \leq \varepsilon,$$

which proves the claim.

(III) It will be shown that Λ is completely continuous. Define the set $B_r(Z)$ for the closed ball with center at 0 and radius r in the space Z. Let $V = \Lambda(B_r(C_{\hbar}(X)))$ and $v = \Lambda(u)$ for $u \in B_r(C_{\hbar}(X))$. Firstly, we will prove that $V(t)$ is a relatively compact subset of X for each $t \in \mathbb{R}$. It follows from the condition (Cf3) that the function $s \to \Phi(s)W_f(r\hbar(t - s))$ is integrable on $\mathbb{R}_+$. Hence, for $\varepsilon > 0$, we can choose $a > 0$ such that $\int_a^{\infty} \Phi(s)W_f(r\hbar(t-s))ds \leq \varepsilon$. Since for any $0 < \tau < a$

$$v_\tau(t) = \int_\tau^a S_{\alpha,\beta}(s)f(t - s, u(t - s)) \, ds$$
$$+ \int_a^{\infty} S_{\alpha,\beta}(s)f(t - s, u(t - s)) \, ds$$

and

$$\left\| \int_a^{\infty} S_{\alpha,\beta}(s)f(t - s, u(t - s)) \, ds \right\| \leq \int_a^{\infty} \Phi(s)W_f(r\hbar(t - s))ds \leq \varepsilon,$$

we get $v_\tau(t) \in (a-\tau)\overline{co(K)} + B_\varepsilon(X)$ (see Lemma 1.7), where $co(K)$ denotes the convex hull of K and $K = \{S_{\alpha,\beta}(s)f(\xi, u) : \tau \leq s \leq a, \ t - a \leq \xi \leq t - \tau, \ \|u\|_{\hbar} \leq r\}$. By the condition (H$_C$), we infer that K is a relatively compact set, and $V_\tau(t) \subseteq (a - \tau)\overline{co(K)} + B_\varepsilon(X)$ is also relatively compact for any $\tau > 0$. Since

$$\|v(t) - v_\tau(t)\| \leq \int_0^{\tau} \Phi(s)W_f(r\hbar(t - s))ds \to 0$$

as $\tau \to 0$ by the Lebesgue dominated convergence theorem. Thus, there are relatively compact sets arbitrarily close to the set $V(t)$, which establishes our assertion.

Secondly, we show that the set V is equicontinuous. In fact, we can decompose

$$v(t + s) - v(t)$$
$$= \int_0^{s} S_{\alpha,\beta}(\sigma)f(t + s - \sigma, u(t + s - \sigma)) \, d\sigma$$

$$+ \int_0^a [S_{\alpha,\beta}(\sigma + s) - S_{\alpha,\beta}(\sigma)] f(t - \sigma, u(t - \sigma)) \, d\sigma$$

$$+ \int_a^\infty [S_{\alpha,\beta}(\sigma + s) - S_{\alpha,\beta}(\sigma)] f(t - \sigma, u(t - \sigma)) \, d\sigma.$$

For each $\varepsilon > 0$, we can choose $a > 0$ and $\delta_1 > 0$ such that

$$\left\| \int_0^s S_{\alpha,\beta}(\sigma) f(t + s - \sigma, u(t + s - \sigma)) \, d\sigma \right.$$

$$\left. + \int_a^\infty [S_{\alpha,\beta}(\sigma + s) - S_{\alpha,\beta}(\sigma)] f(t - \sigma, u(t - \sigma)) \, d\sigma \right\|$$

$$\leq \int_0^s \Phi(\sigma) W_f(r\mathrm{h}(t + s - \sigma)) d\sigma$$

$$+ \int_a^\infty [\Phi(\sigma + s) + \Phi(\sigma)] W_f(r\mathrm{h}(t - \sigma)) d\sigma$$

$$\leq \frac{\varepsilon}{2}$$

for $s \leq \delta_1$. Moreover, since $S_{\alpha,\beta}(t), t > 0$ is compact (see (H_C) and Remark 4.5), we can choose a suitable $\widetilde{M} > 0$ such that $\|S_{\alpha,\beta}(t)\| \leq \widetilde{M}$ on $[0, a]$ for each $a > 0$ via (4.3), and

$$\int_0^a \|[S_{\alpha,\beta}(\sigma + s) - S_{\alpha,\beta}(\sigma)] f(t - \sigma, u(t - \sigma))\| d\sigma$$

$$\leq \int_0^a \|[S_{\alpha,\beta}(\sigma + s) - S_{\alpha,\beta}(\sigma)]\| W_f(r\mathrm{h}(t - \sigma)) d\sigma.$$

By the condition (Cf3), we have

$$\|[S_{\alpha,\beta}(\cdot + s) - S_{\alpha,\beta}(\cdot)]\| W_f(r\mathrm{h}(t - \cdot))$$

$$\leq 2\widetilde{M} W_f(r\mathrm{h}(t - \cdot)) \in L^1([0, a], \mathbb{R}).$$

Since the function $t \mapsto S_{\alpha,\beta}(t)$ is norm continuous, see Ponce (2016, Proposition 11) (or Theorem A.1 in Appendix A), $S_{\alpha,\beta}(\sigma + s) - S_{\alpha,\beta}(\sigma) \to 0$ in $\mathcal{B}(X)$ as $s \to 0$. By the Lebesgue's dominated convergence theorem we can choose $\delta_2 > 0$ such that

$$\left\| \int_0^a [S_{\alpha,\beta}(\sigma + s) - S_{\alpha,\beta}(\sigma)] f(t - \sigma, u(t - \sigma)) \, d\sigma \right\| \leq \frac{\varepsilon}{2}$$

for $s \leq \delta_2$. Combining these estimates, we get $\|v(t + s) - v(t)\| \leq \varepsilon$ for s sufficiently small and independent of $u \in B_r(C_{\mathrm{h}}(X))$.

Finally, from the condition (Cf3), it is easy to see that

$$\frac{\|v(t)\|}{\hbar(t)} \le \frac{1}{\hbar(t)} \int_{-\infty}^{t} \Phi(t-s) W_f(r\hbar(s)) ds \to 0, \quad |t| \to \infty,$$

and this convergence is independent of $u \in B_r(C_{\hbar}(X))$. Hence, by Lemma 1.20 of Chapter 1, V is a relatively compact set in $C_{\hbar}(X)$.

(IV) Let $u^{\lambda}(\cdot)$ be a solution to $u^{\lambda} = \lambda \Lambda(u^{\lambda})$ for some $0 < \lambda < 1$. We can estimate

$$\|u^{\lambda}(t)\| = \lambda \left\| \int_{-\infty}^{t} S_{\alpha,\beta}(t-s) f\left(s, u^{\lambda}(s)\right) ds \right\|$$

$$\le \int_{-\infty}^{t} \Phi(t-s) W_f(\|u^{\lambda}\|_{\hbar} \hbar(s)) ds$$

$$\le \beta(\|u^{\lambda}\|_{\hbar}) \hbar(t).$$

Hence, we have

$$\frac{\|u^{\lambda}\|_{\hbar}}{\beta(\|u^{\lambda}\|_{\hbar})} \le 1,$$

and combining with the condition (Cf5), we conclude that the set $\{u^{\lambda} : u^{\lambda} = \lambda \Lambda(u^{\lambda}), \ \lambda \in (0,1)\}$ is bounded.

(V) From Corollary 2.8(1) of Chapter 2 and Theorem 4.4, $\Lambda(PBP_{\omega,k}(\mathbb{R}, X)) \subseteq PBP_{\omega,k}(\mathbb{R}, X)$. It is noticed that $\overline{PBP_{\omega,k}(\mathbb{R}, X)}$ is a closed subspace of $C_{\hbar}(X)$, thus we can consider $\Lambda : \overline{PBP_{\omega,k}(\mathbb{R}, X)} \to \overline{PBP_{\omega,k}(\mathbb{R}, X)}$. It is deduced from steps (I)–(III) that Λ is completely continuous. From Lemma 1.22 of Chapter 1 we infer that Λ has a fixed point $u \in \overline{PBP_{\omega,k}(\mathbb{R}, X)}$. Let $\{u_n\}$ be a sequence in $PBP_{\omega,k}(\mathbb{R}, X)$ which converges to u by the norm of $C_{\hbar}(X)$, we have

$$\|\Lambda u^n - u\|_{\infty} = \|\Lambda u^n - \Lambda u\|_{\infty}$$

$$\le \sup_{t \in \mathbb{R}} \frac{1}{\hbar(t)} \int_{0}^{t} \Phi(t-s) \| f\left(s, u^n(s)\right) - f(s, u(s)) \| ds.$$

It follows from (Cf4) that $\Lambda u^n \to u$, as $n \to \infty$ uniformly in $\mathbb{R}$. Since $\Lambda u^n \in PBP_{\omega,k}(\mathbb{R}, X)$, we have $u \in PBP_{\omega,k}(\mathbb{R}, X)$, i.e., Eq. (4.1) admits one mild solution $u \in PBP_{\omega,k}(\mathbb{R}, X)$. $\qquad \square$

Remark 4.6. Let $\rho \in \mathcal{U}_{\text{inv}}$ and take into account Theorem 4.4, Corollary 2.8(2) of Chapter 2, we can obtain Eq. (4.1) admits at least one mild solution $u \in WPBP_{\omega,k}(\mathbb{R}, X, \rho)$ such as Theorem 4.8.

Theorem 4.9. *Let conditions* (H_A), (H_C) *and* $(Cf1)$–$(Cf5)$ *hold. Assume that* $f \in BC(\mathbb{R} \times X, X)$ *satisfies* $(T1)$ *in Theorem 2.5 of Chapter 2. Then Eq. (4.1) admits at least one mild solution* $u \in SABP_{\omega,k}(\mathbb{R}, X)$.

Proof. Taking into account Theorems 2.6 of Chapter 2 and 4.4, we can complete the proof similarly to that of Theorem 4.8. We omit details here.

$\square$

Remark 4.7. Taking into account Theorems 2.9, 2.12 of Chapter 2, and 4.4, we can obtain Eq. (4.1) admits at least one mild solution $u \in PSABP_{\omega,k}(\mathbb{R}, X, \rho)$ (or $WPSABP_{\omega,k}(\mathbb{R}, X, \rho)$) such as Theorems 4.8, respectively.

Example 4.1. Let $\tau < 0$, $1 < \alpha < 2$ and $1 \leq \beta < \alpha$. Take $X := L^2[0, \pi]$ and consider the problem

$$\begin{cases} \partial_t^\alpha u(x,t) = \dfrac{\partial^2}{\partial x^2} u(x,t) + \tau u(x,t) + \varepsilon \partial_t^{\alpha-\beta} f(t, u(x,t)) \\ u(t,0) = u(t,\pi) = 0, \ t \in \mathbb{R}, \end{cases} \tag{4.11}$$

where $(x,t) \in [0, 2\pi] \times \mathbb{R}$, $\varepsilon > 0$ and $\phi \in X$. Consider the operator

$$A := \frac{\partial^2}{\partial x^2} u + \tau u$$

on X with domain

$$D(A) := \{u \in L^2[0, \pi] : u'' \in L^2[0, \pi]; u(0) = u(\pi) = 0\}.$$

Then, A is ϖ-sectorial with $\varpi = \tau < 0$ of angle $\pi/2$ (and hence of angle $(\alpha - 1)\pi/2$ for all $1 < \alpha < 2$). Therefore, A generates a resolvent operator family $\{S_{\alpha,\beta}\}_{t \geq 0}$ by Theorem 4.2.

Let $f(t, \phi)(s) := g(t, \phi)(s) + h(t, \phi)(s)$ with

$$g(t, \phi)(s) := \gamma(t)\varrho(\phi(s)), \ h(t, \phi)(s) := e^{-t^2} \cos(\phi(s)).$$

Assume that $\gamma(t)$ is a bounded continuous T-periodic function, i.e., $\gamma(t + T) = \gamma(t)$, and ϱ satisfies the condition $\varrho(e^{ikT}x) = e^{ikT}\varrho(x), \|\varrho(x) - \varrho(y)\|_X \leq L_\varrho \|x - y\|_X$, $L_\varrho > 0$. Then we have $g(t + T, e^{ikT}\phi)(s) = \gamma(t + T)\varrho(e^{ikT}\phi(s)) = \gamma(t)e^{ikT}\varrho(\phi(s)) = e^{ikT}g(t, \phi)(s)$. Meanwhile, it is easy to see that $h(t, \phi) \in \mathscr{E}(\mathbb{R} \times X, X)$. We also have

$$\|f(t, \phi) - f(t, \psi)\|_{L^2[0,\pi]}^2 \leq \left[|\gamma(t)|^2 + 1\right] L_\varrho^2 \|\phi(s) - \psi(s)\|_{L^2[0,\pi]}^2.$$

Suppose further that $|\gamma(t)| \in L^1(\mathbb{R}, \mathbb{R}_+)$, then by Theorem 4.5, the problem (4.11) admits a unique pseudo Bloch-type periodic mild solution on $\mathbb{R}$ for small enough ε.

Chapter 5

Bloch-Type Periodic Solutions to Fractional Evolution Equations of Sobolev Type

In this chapter, we consider the existence and uniqueness of generalized Bloch-type periodic mild solutions to the following semilinear fractional differential equation of Sobolev type

$$\partial_t^\alpha (Eu)(t) = Au(t) + \partial_t^{\alpha-\beta}(Ef)(t, u(t)), \quad t \in \mathbb{R}, \qquad (5.1)$$

where A and E are closed linear operators defined on a Banach space X, $1 \leq \beta < \alpha < 2$, $f \in BC(\mathbb{R} \times X, [D(E)])$ and ∂_t^α denotes the Weyl fractional derivative.

The existence of mild solutions to fractional differential equations of Sobolev type has been studied in the last years, see for instance Chang *et al.* (2017), Chang and Ponce (2019), Chang *et al.* (2019a), Chang and Pei (2020), Debbouche and Nieto (2014), Debbouche and Torres (2015), Li *et al.* (2012), Liang *et al.* (2019), Lizama and Ponce (2011a), Pei and Chang (2020), Ponce (2014, 2017, 2020d), Zhao and Chang (2020) and references therein. Sobolev-type differential equations describes several partial differential equations arising in physics and applied sciences. For example, if $A = \Delta$ is the Laplacian and $E = m$ is the multiplication operator by a function $m(x)$, then model in the form of (5.1) describes the infiltration of water in unsaturated porous media. See for instance Carroll and Showalter (1976) and Favini and Yagi (1999) for further details.

It is noted that Eq. (5.1) has been treated in some cases. For instance, if $1 < \alpha < 2, \beta = 1$, A is a sectorial operator and $E = I$ (the identity operator on X), then the following equation:

$$\partial_t^\alpha u(t) = Au(t) + \partial_t^{\alpha-1} f(t, u(t)), \quad t \in \mathbb{R},$$

has been widely studied through a resolvent family $\{R_\alpha(t)\}_{t \geq 0}$ with its Laplace transform $\widehat{R}_\alpha(\lambda) = \lambda^{\alpha-1}(\lambda^\alpha - A)^{-1}$, which is generated by A and decays in norm as $\frac{1}{1+|\varpi|t^\alpha}$ (see Cuesta, 2007). We can refer to Araya and

Lizama (2008), Chang *et al.* (2012), Chang and Wei (2021b), Cuevas and Lizama (2008), Cuevas and Souza (2009), Mophou (2011), Ponce (2013b), Wei and Chang (2022) and references cited therein for detailed results.

Chang *et al.* (2019b) introduced a Sobolev-type resolvent family (also called characteristic solution operators) $\{S^E_{\alpha,\beta}(t)\}_{t\geq 0}$ which allows to write out the formulation of mild solutions to Eq. (5.1). By studying the asymptotic decay of the Sobolev-type resolvent family $\{S^E_{\alpha,\beta}(t)\}_{t\geq 0}$, authors also established the existence and uniqueness of (asymptotically) almost periodic and almost automorphic mild solutions to Eq. (5.1). It is not necessarily assumed the existence or compactness of the inverse E^{-1} as well as any assumption on the inclusion relation between $D(A)$ and $D(E)$ in Chang *et al.* (2019b). In this chapter, we continue to investigate the existence and uniqueness of generalized Bloch-type periodic mild solutions to Eq. (5.1) as an application.

5.1   Asymptotic Behavior of Sobolev-Type Resolvent Family

In this section, we recall some results on asymptotic behavior of Sobolev-type resolvent family which was established in Chang *et al.* (2019b).

A closed operator A, defined on a Banach space X, is said to be ϖ-*sectorial with respect to E of angle* ϕ (or *the pair (A, E) is ϖ-sectorial of angle ϕ*), if there exist $\phi \in [0, \pi/2)$ and $\varpi \in \mathbb{R}$ such that its *E-resolvent operator* $(\lambda E - A)^{-1}$ exists in the sector

$$\varpi + \Sigma_\phi := \left\{\varpi + \lambda : \lambda \in \mathbb{C}, |\arg(\lambda)| < \frac{\pi}{2} + \phi\right\} \setminus \{\varpi\}$$

and

$$\|(\lambda E - A)^{-1}E\| \leq \frac{K}{|\lambda - \varpi|}, \quad \lambda \in \varpi + \Sigma_\phi.$$

A class of such operators is the operator A which is 0-sectorial with respect to E, see Sviridyuk and Fedorov (2003, Chapter 3).

Definition 5.1. Let A, E be closed and linear operators with domain $D(A) \cap D(E) \neq \{0\}$ defined on a Banach space X, and $\alpha, \beta > 0$. We say that the pair (A, E) is the generator of an (α, β)-resolvent family, if there exist $\widetilde{\omega} \geq 0$ and a strongly continuous function $S^E_{\alpha,\beta} : [0, \infty) \to \mathcal{B}([D(E)], X)$ such that $\{\lambda^\alpha : \text{Re}\lambda > \widetilde{\omega}\} \subset \rho_E(A)$ and for all $x \in D(E)$,

$$\lambda^{\alpha-\beta}(\lambda^\alpha E - A)^{-1}Ex = \int_0^\infty e^{-\lambda t}S^E_{\alpha,\beta}(t)x\,dt, \quad \text{Re}\lambda > \widetilde{\omega},$$

where

$$\rho_E(A) := \{\mu \in \mathbb{C} : \mu E - A \text{ is invertible and } (\mu E - A)^{-1} \text{ is bounded}\}.$$

In this case, $\{S_{\alpha,\beta}^E(t)\}_{t\geq 0}$ is called the (α, β)-*resolvent family* generated by the pair (A, E).

Remark 5.1. Define the function $g_\alpha(t) = \frac{t^{\alpha-1}}{\Gamma(\alpha)}$ for all $t \geq 0$. It is easy to show from Lemmas 1.16 and 1.17 of Chapter 1 (see also Lizama, 2000, Proposition 3.1, Lemma 2.2) that if (A, E) generates an (α, β)-resolvent family $\{S_{\alpha,\beta}^E(t)\}_{t\geq 0}$, then it satisfies the following properties:

(i) $S_{\alpha,\beta}^E(0)E = g_\alpha(0)E$;
(ii) $(g_\alpha * S_{\alpha,\beta}^E)(t)x \in D(A) \cap D(E)$ and

$$ES_{\alpha,\beta}^E(t)x = g_\beta(t)Ex + A\int_0^t g_\alpha(t-s)S_{\alpha,\beta}^E(s)x\,ds,$$

for all $x \in D(E)$ and $t \geq 0$.

The following result, analogous to the Hille–Yosida theorem for C_0-semigroups, can be obtained similarly to Lemma 1.18 of Chapter 1 (see also Lizama, 2000, Theorem 3.4).

Theorem 5.1. *Let A, E be closed linear operators defined on a Banach space X. Then the following assertions are equivalent:*

(1) *The pair (A, E) generates an (α, β)-resolvent family $\{S_{\alpha,\beta}^E(t)\}_{t\geq 0}$ satisfying $\|S_{\alpha,\beta}^E(t)\| \leq Me^{\mu t}$ for all $t \geq 0$ and for some constants $M > 0$ and $\mu \in \mathbb{R}$.*
(2) *There exist constants $\mu \in \mathbb{R}$ and $M > 0$ such that $\lambda^\alpha \in \rho_E(A)$ for all λ with $\lambda > \mu$ and $H(\lambda) := \lambda^{\alpha-\beta}(\lambda^\alpha E - A)^{-1}E$ satisfies the estimate*

$$\|H^{(n)}(\lambda)\| \leq \frac{Mn!}{(\lambda - \mu)^{n+1}},$$

for all $\lambda > \mu$ and $n \in \mathbb{N} \cup \{0\}$.

The next result gives conditions on operators A and E in order to generate an (α, β)-resolvent family.

Theorem 5.2. *Let $1 < \alpha < 2$ and $\beta \geq 1$ such that $\alpha - \beta + 1 > 0$. Assume that A is a ϖ-sectorial operator with respect to E of angle $0 \leq \phi < (\alpha-1)\frac{\pi}{2}$, where $\varpi < 0$. Then the pair (A, E) generates the (α, β)-resolvent family $\{S_{\alpha,\beta}^E(t)\}_{t\geq 0}$.*

Proof. For $\lambda = re^{i\theta}$ with $|\theta| < \pi/2$ and $r > 0$, we define $g(\lambda) = \lambda^\alpha$. We observe that

$$\arg\left(g\left(re^{i\theta}\right)\right) = \operatorname{Im}\log\left(g\left(re^{i\theta}\right)\right) = \operatorname{Im}\int_0^\theta \frac{d}{dt}\log\left(g\left(re^{it}\right)\right)dt$$

$$= \operatorname{Im}\int_0^\theta \frac{g'(re^{it})ire^{it}}{g(re^{it})}dt$$

with

$$\frac{\lambda g'(\lambda)}{g(\lambda)} = \alpha.$$

Therefore

$$\left|\arg\left(g\left(re^{i\theta}\right)\right)\right| \leq \alpha|\theta| < (\alpha - 1)\frac{\pi}{2} + \frac{\pi}{2}.$$

We conclude that $\lambda^\alpha \in \Sigma_{(\alpha-1)\frac{\pi}{2}}$ for all $\operatorname{Re}\lambda > 0$. From the above, we have that $H(\lambda) = \lambda^{\alpha-\beta}(\lambda^\alpha E - A)^{-1}E$ is well defined and satisfies

$$\|\lambda H(\lambda)\| \leq \frac{K|\lambda|^{\alpha-\beta+1}}{|\lambda^\alpha - \varpi|} \leq M_1 \quad \text{for all} \ \ \operatorname{Re}(\lambda) > 0,$$

where M_1 is a positive constant. On the other hand,

$$\|\lambda^2 H'(\lambda)\| \leq |\alpha - \beta|\,\|\lambda H(\lambda)\| + \alpha\|\lambda H(\lambda)\|\,\left\|\lambda^\alpha(\lambda^\alpha E - A)^{-1}E\right\|$$

$$\leq |\alpha - \beta|\,\|\lambda H(\lambda)\| + \alpha\|\lambda H(\lambda)\|\frac{|\lambda^\alpha|}{|\lambda^\alpha - \varpi|}$$

$$\leq M_2,$$

for all $\operatorname{Re}\lambda > 0$ and a constant $M_2 > 0$. By using Lemma 1.10 of Chapter 1 (cf. Prüss, 1993, Proposition 0.1) and Theorem 5.1, we obtain that the pair (A, E) generates a resolvent family $\{S^E_{\alpha,\beta}(t)\}_{t\geq 0}$. $\qquad\square$

The following result is involved in asymptotic behavior of the resolvent family $\{S^E_{\alpha,\beta}(t)\}_{t\geq 0}$.

Theorem 5.3. *Let $1 < \alpha < 2$ and $\beta \geq 1$ such that $\alpha - \beta + 1 > 0$. Let A and E be closed linear operators on X, $D(A) \cap D(E) \neq \{0\}$. Suppose A is a ϖ-sectorial operator with respect to E of angle $0 \leq \phi < (\alpha - 1)\frac{\pi}{2}$, where $\varpi < 0$. Then, there exists a constant $M > 0$ depending only upon α and β such that the resolvent family $\{S^E_{\alpha,\beta}(t)\}_{t\geq 0}$ generated by the pair (A, E) satisfies*

$$\left\|S^E_{\alpha,\beta}(t)\right\| \leq \frac{Mt^{\beta-1}}{1 + |\varpi|t^\alpha}, \quad \text{for all} \ \ t > 0. \tag{5.2}$$

Proof. We exploit some ideas of Keyantuo *et al.* (2013). Since A is ϖ-sectorial with respect to E of angle $0 \leq \phi < (\alpha - 1)\frac{\pi}{2}$, we have by Theorem 5.2 that the pair (A, E) generates an (α, β)-resolvent family $\{S^E_{\alpha,\beta}(t)\}_{t \geq 0}$. As in the proof of Theorem 5.2 we have that $\lambda^\alpha \in \rho_E(A)$ for all $\lambda^\alpha \in \Sigma_\phi$. Moreover, the Laplace transform of $S^E_{\alpha,\beta}(t)$ satisfies

$$\widehat{S^E_{\alpha,\beta}}(\lambda) = \lambda^{\alpha-\beta}(\lambda^\alpha E - A)^{-1}E$$

for all $\lambda^\alpha \in \rho_E(A)$. The inversion theorem of the Laplace transform implies

$$S^E_{\alpha,\beta}(t) = \frac{1}{2\pi i}\int_\gamma e^{\lambda t}\lambda^{\alpha-\beta}(\lambda^\alpha E - A)^{-1}E d\lambda, \tag{5.3}$$

where γ is a suitable positively oriented path. We define γ as the path whose support Γ is given by

$$\Gamma := \{\lambda : \lambda \in \mathbb{C}, \lambda^\alpha \text{ belongs to the boundary of } B_{\frac{1}{t^\alpha}}, t > 0\},$$

where $B_{\frac{1}{t^\alpha}}$ is given by

$$B_{\frac{1}{t^\alpha}} := \left\{\frac{1}{t^\alpha} + \Sigma_\theta\right\} \cup \{\varpi + \Sigma_\phi\},$$

and $\phi < \theta < \frac{\pi}{2}$. Note that, with this path γ the function $S^E_{\alpha,\beta}(t)$ given in (5.3), is well defined.

Since A is ϖ-sectorial with respect to E of angle $0 \leq \phi < (\alpha - 1)\frac{\pi}{2}$, it follows that

$$\|(\lambda E - A)^{-1}E\| \leq \frac{K}{|\lambda - \varpi|},$$

for all $\lambda \in \mathbb{C}$ with $\lambda \in \varpi + \Sigma_\phi, \lambda \neq \varpi$.

Now, we split γ into two paths, γ_1, γ_2, whose supports Γ_1 and Γ_2 are given by

$$\Gamma_1 := \Gamma \cap \overline{\left\{\frac{1}{t^\alpha} + \Sigma_\theta\right\}} \quad \text{and} \quad \Gamma_2 = \Gamma \cap \overline{\{\varpi + \Sigma_\phi\}}.$$

Therefore, $\Gamma = \Gamma_1 \cup \Gamma_2$ and $S^E_{\alpha,\beta}(t) = I_1(t) + I_2(t)$, where

$$I_j(t) := \frac{1}{2\pi i}\int_{\gamma_j} e^{\lambda t}\lambda^{\alpha-\beta}(\lambda^\alpha E - A)^{-1}E d\lambda, \quad j = 1, 2.$$

First, we estimate $\|I_1(t)\|$. We define $\lambda_{\min}$ as the complex $\lambda \in \mathbb{C}$ such that $\text{Im}(\lambda) > 0$, and $|\lambda^\alpha_{\min} - \varpi| = \text{dist}(L, \varpi)$, where L in the line passing by

$\left(\frac{1}{t^\alpha}, 0\right)$ and the intersection of Γ_1 and Γ_2. For $\lambda \in \Gamma_1$ and $\varpi < 0$ we have that

$$\cos(\theta) = \sin\left(\frac{\pi}{2} - \theta\right) = \frac{|\lambda_{\min}^\alpha - \varpi|}{|\varpi| + \frac{1}{t^\alpha}} \leq \frac{|\lambda^\alpha - \varpi|}{|\varpi| + \frac{1}{t^\alpha}}.$$

Therefore, if $\lambda \in \Gamma_1$ then

$$\frac{1}{|\lambda^\alpha - \varpi|} \leq \frac{t^\alpha}{\cos(\theta)(1 + |\varpi|t^\alpha)}.$$

Hence,

$$\|I_1(t)\| \leq \frac{K}{2\pi} \frac{t^\alpha}{\cos(\theta)(1 + t^\alpha|\varpi|)} \int_{\gamma_1} |e^{\lambda t}| |\lambda|^{\alpha - \beta} |d\lambda|$$

$$\leq \frac{K}{\pi} \frac{t^\alpha}{\cos(\theta)(1 + t^\alpha|\varpi|)} \int_0^\infty e^{-t\sin(\theta)s} s^{(\alpha-\beta)} ds$$

$$= \frac{K}{\pi} \frac{t^\alpha}{\cos(\theta)(1 + t^\alpha|\varpi|)} \frac{\Gamma(\alpha - \beta + 1)}{\sin(\theta)^{\alpha-\beta+1} t^{\alpha-\beta+1}}$$

$$\leq \frac{M_1 t^{\beta-1}}{1 + t^\alpha|\varpi|}.$$

Next, we estimate $\|I_2(t)\|$. Let z_t be the intersection point between the boundary of $\frac{1}{t^\alpha} + \Sigma_\theta$ and $\varpi + \Sigma_\phi$. We notice that for all $\lambda \in \Gamma_2$, we have (by using the law of sines)

$$|z_t - \varpi| = \frac{|\varpi| + \frac{1}{t^\alpha}}{\sin(\theta - \phi)} \cos(\theta), \quad t > 0.$$

Hence, if $\lambda \in \Gamma_2$, then

$$\frac{1}{|\lambda^\alpha - \varpi|} \leq \frac{\sin(\theta - \phi)}{\cos(\theta)\cos(\phi)} \frac{t^\alpha}{(1 + |\varpi|t^\alpha)}.$$

Thus, there exists a constant $C > 0$ (depending only on θ and ϕ) such that

$$\|I_2(t)\| \leq \frac{KC}{2\pi} \frac{t^\alpha}{(1 + t^\alpha|\varpi|)} \int_{\gamma_2} |e^{\lambda t}| |\lambda|^{\alpha - \beta} |d\lambda|$$

$$\leq \frac{KC}{\pi} \frac{t^\alpha}{(1 + t^\alpha|\varpi|)} \int_0^\infty e^{-t\sin(\phi)s} s^{(\alpha-\beta)} ds$$

$$= \frac{KC}{\pi} \frac{t^\alpha}{(1 + t^\alpha|\varpi|)} \frac{\Gamma(\alpha - \beta + 1)}{\sin(\phi)^{\alpha-\beta+1} t^{\alpha-\beta+1}}$$

$$\leq \frac{M_2 t^{\beta-1}}{1 + t^\alpha|\varpi|}.$$

Therefore, there exists a constant M depending only on α and β such that

$$\|S_{\alpha,\beta}^{E}(t)\| \leq \|I_1(t)\| + \|I_2(t)\| \leq \frac{Mt^{\beta-1}}{1 + |\varpi|t^{\alpha}}, \quad t > 0.$$

This ends the proof. $\qquad\qquad\qquad\qquad\qquad\qquad\qquad\qquad\qquad\square$

From Theorem 5.3, we have the following lemma.

Lemma 5.1. *If* $1 \leq \beta < \alpha < 2$ *and the pair* (A, E) *is a* ϖ-*sectorial operator of angle* $0 \leq \phi < (\alpha - 1)\frac{\pi}{2}$, *where* $\varpi < 0$, *then* $\{S_{\alpha,\beta}^{E}(t)\}_{t\geq 0}$ *is uniformly integrable.*

Proof. The condition $1 \leq \beta < \alpha < 2$ implies $\alpha - \beta + 1 > 0$ and from (5.2) we have

$$\int_0^{\infty} \|S_{\alpha,\beta}^{E}(t)\| dt \leq M \int_0^{\infty} \frac{t^{\beta-1}}{1 + |\varpi|t^{\alpha}} dt$$

$$= M \frac{\left(\frac{1}{|\varpi|}\right)^{\beta-1/\alpha}}{\alpha|\varpi|^{1/\alpha}} \int_0^{\infty} \frac{u^{(\beta-\alpha)/\alpha}}{1 + u} du$$

$$= \frac{M}{\alpha|\varpi|^{\beta/\alpha}} \mathbf{B}\left(\frac{\beta}{\alpha}, 1 - \frac{\beta}{\alpha}\right),$$

where $\mathbf{B}(\cdot, \cdot)$ denotes the Beta function. Since $\frac{\beta}{\alpha} > 0$ and $1 - \frac{\beta}{\alpha} > 0$, then we have the assertion. $\qquad\qquad\qquad\qquad\qquad\qquad\qquad\qquad\square$

5.2 Bloch-Type Periodic Solutions

We first consider the following Sobolev-type linear fractional differential equation

$$\partial_t^{\alpha}(Eu)(t) = Au(t) + \partial_t^{\alpha-\beta}(Ef)(t), \quad t \in \mathbb{R}. \qquad (5.4)$$

Assume that the pair (A, E) is the generator of an (α, β)-resolvent family $\{S_{\alpha,\beta}^{E}(t)\}_{t\geq 0}$ which is uniformly integrable, i.e.,

$$\int_0^{\infty} \|S_{\alpha,\beta}^{E}(t)\| dt < \infty.$$

For a given $f \in BC(\mathbb{R}, X)$, define the function $\Phi(t)$ by

$$\Phi(t) := \int_{-\infty}^{t} S_{\alpha,\beta}^{E}(t - s)f(s)ds, \quad t \in \mathbb{R}.$$

If $f(t) \in D(E)$ for all $t \in \mathbb{R}$, then $\Phi(t) \in D(E)$ for all $t \in \mathbb{R}$, see Lemma 1.4 (or Arendt *et al.*, 2001, Proposition 1.1.7). Take $n = [\alpha] + 1$ and assume the existence of $\partial_t^\alpha(E\Phi)$. From the Fubini's theorem we get

$$\partial_t^\alpha(E\Phi)(t)$$

$$= \frac{d^n}{dt^n}\int_{-\infty}^{t} g_{n-\alpha}(t-s)E\Phi(s)ds$$

$$= \frac{d^n}{dt^n}\int_{-\infty}^{t} g_{n-\alpha}(t-s)\int_{-\infty}^{s} ES_{\alpha,\beta}^{E}(s-r)f(r)drds$$

$$= \frac{d^n}{dt^n}\int_{-\infty}^{t} g_{n-\alpha}(t-s)$$

$$\times \int_{-\infty}^{s}\left[g_\beta(s-r)Ef(r) + A(g_\alpha * S_{\alpha,\beta}^{E})(s-r)f(r)\right]drds$$

$$= \frac{d^n}{dt^n}\int_{-\infty}^{t} g_{n-\alpha}(t-s)\partial_t^{-\beta}Ef(s)ds + \frac{d^n}{dt^n}\int_{-\infty}^{t} g_{n-\alpha}(t-s)$$

$$\times \int_{-\infty}^{s} A\int_{0}^{s-r} g_\alpha(s-r-v)S_{\alpha,\beta}^{E}(v)f(r)dvdrds$$

$$= \partial_t^{\alpha-\beta}(Ef)(t) + \frac{d^n}{dt^n}\int_{-\infty}^{t} g_{n-\alpha}(t-s)$$

$$\times \int_{-\infty}^{s} A\int_{r}^{s} g_\alpha(s-w)S_{\alpha,\beta}^{E}(w-r)f(r)dwdrds$$

$$= \partial_t^{\alpha-\beta}(Ef)(t) + \frac{d^n}{dt^n}\int_{-\infty}^{t} g_{n-\alpha}(t-s)$$

$$\times \int_{-\infty}^{s} A\int_{-\infty}^{w} g_\alpha(s-w)S_{\alpha,\beta}^{E}(w-r)f(r)drdwds$$

$$= \partial_t^{\alpha-\beta}(Ef)(t) + \frac{d^n}{dt^n}\int_{-\infty}^{t} g_{n-\alpha}(t-s)$$

$$\times \int_{-\infty}^{s} g_\alpha(s-w)A\Phi(w)dwds = \partial_t^{\alpha-\beta}(Ef)(t) + A\Phi(t),$$

for all $t \in \mathbb{R}$, which means that Φ is a solution to Eq. (5.4). Generally, we have only $\Phi(t) \in X$ or $\partial_t^\alpha(E\Phi)$ does not exist. We introduce the following definition of solution.

Definition 5.2. Assume that the pair (A, E) generates an (α, β)-resolvent family $\{S_{\alpha,\beta}^E(t)\}_{t \geq 0}$. A continuous function $u \in C(\mathbb{R}, X)$ is said to be a mild solution to Eq. (5.4) if the function $s \mapsto S_{\alpha,\beta}^E(t - s)f(s)$ is integrable on $(-\infty, t)$ for each $t \in \mathbb{R}$ and

$$u(t) = \int_{-\infty}^{t} S_{\alpha,\beta}^E(t - s)f(s)ds, \quad t \in \mathbb{R}.$$

Theorem 5.4. *Let $1 \leq \beta < \alpha < 2$ and suppose further that the pair (A, E) is a ϖ-sectorial operator of angle $0 \leq \phi < (\alpha - 1)\frac{\pi}{2}$, where $\varpi < 0$. If $f \in \mathcal{N}([D(E)])$, then Eq. (5.4) has a unique mild solution $u \in \mathcal{N}([D(E)])$.*

Proof. It follows from Theorem 5.2 and Lemma 5.1 that the (α, β)-resolvent family $\{S_{\alpha,\beta}^E(t)\}_{t \geq 0}$ generated by the pair (A, E) is uniformly integrable. For a given $f \in \mathcal{N}([D(E)])$, we can conclude that the mild solution $u \in \mathcal{N}([D(E)])$ combined with Lemma 1.4 (or Arendt *et al.*, 2001, Proposition 1.1.7) and Theorems 2.1, 2.2, 2.7, 2.10, 2.13 of Chapter 2, respectively. The uniqueness is easily obtained. $\quad\square$

Definition 5.3. Assume that the pair (A, E) generates an (α, β)-resolvent family $\{S_{\alpha,\beta}^E(t)\}_{t \geq 0}$. A function $u \in C(\mathbb{R}, X)$ is said to be a mild solution to Eq. (5.1) if the function $s \mapsto S_{\alpha,\beta}^E(t - s)f(s, u(s))$ is integrable on $(-\infty, t)$ for each $t \in \mathbb{R}$ and

$$u(t) = \int_{-\infty}^{t} S_{\alpha,\beta}^E(t - s)f(s, u(s))ds, \quad t \in \mathbb{R}.$$

Theorem 5.5. *Let $1 \leq \beta < \alpha < 2$ and assume that the pair (A, E) is a ϖ-sectorial operator of angle $0 \leq \phi < (\alpha - 1)\frac{\pi}{2}$, where $\varpi < 0$. Suppose further that $f = g + h \in BC(\mathbb{R} \times X, [D(E)])$ with g satisfying the condition (A1) in Section 2.3 and $h \in \mathscr{E}(\mathbb{R} \times X, [D(E)])$. If there exists a constant*

$$0 < L < \frac{\alpha}{M}|\varpi|^{\beta/\alpha}\mathbf{B}\left(\frac{\beta}{\alpha}, 1 - \frac{\beta}{\alpha}\right)^{-1},$$

where M is the constant given in Theorem 5.3, and $\mathbf{B}(\cdot, \cdot)$ denotes the Beta function, such that

$$\|f(t, x) - f(t, y)\| \leq L\|x - y\|, \text{ for all } t \in \mathbb{R}, \text{ and } x, y \in X, \quad (5.5)$$

then Eq. (5.1) has a unique mild solution $u \in PBP_{\omega,k}(\mathbb{R}, [D(E)])$.

Proof. Define $F : PBP_{\omega,k}(\mathbb{R}, [D(E)]) \to PBP_{\omega,k}(\mathbb{R}, [D(E)])$ by

$$(F\Phi)(t) := \int_{-\infty}^{t} S_{\alpha,\beta}^{E}(t-s)f(s,\Phi(s))\,ds, \quad t \in \mathbb{R}. \tag{5.6}$$

For each $\Phi \in PBP_{\omega,k}(\mathbb{R}, [D(E)])$, we obtain from Corollary 2.2(1) of Chapter 2, that the function $s \mapsto f(s,\Phi(s))$ belongs to $PBP_{\omega,k}(\mathbb{R}, [D(E)])$. It follows from Theorem 5.4 that $F\Phi \in PBP_{\omega,k}(\mathbb{R}, [D(E)])$ and F is well defined. For $\Phi_1, \Phi_2 \in PBP_{\omega,k}(\mathbb{R}, [D(E)])$ and $t \in \mathbb{R}$, we have:

$$\|(F\Phi_1)(t) - (F\Phi_2)(t)\|$$

$$\leq \int_{-\infty}^{t} \|S_{\alpha,\beta}^{E}(t-s)[f(s,\Phi_1(s)) - f(s,\Phi_2(s))]\|ds$$

$$\leq \int_{-\infty}^{t} L\|S_{\alpha,\beta}^{E}(t-s)\| \cdot \|\Phi_1(s) - \Phi_2(s)\|ds$$

$$\leq L\|\Phi_1 - \Phi_2\|_{\infty} \int_{0}^{\infty} \|S_{\alpha,\beta}^{E}(r)\|dr$$

$$\leq \frac{LM}{\alpha}|\varpi|^{-\beta/\alpha}\mathbf{B}\left(\frac{\beta}{\alpha}, 1 - \frac{\beta}{\alpha}\right)\|\Phi_1 - \Phi_2\|_{\infty},$$

i.e.,

$$\|F\Phi_1 - F\Phi_2\|_{\infty} \leq \frac{LM}{\alpha}|\varpi|^{-\beta/\alpha}\mathbf{B}\left(\frac{\beta}{\alpha}, 1 - \frac{\beta}{\alpha}\right)\|\Phi_1 - \Phi_2\|_{\infty}.$$

This shows that F is a contraction, and by the Banach fixed point theorem there exists a unique $u \in PBP_{\omega,k}(\mathbb{R}, [D(E)])$ such that $Fu = u$. □

Theorem 5.6. *Let $1 \leq \beta < \alpha < 2$ and assume that the pair (A, E) is a ϖ-sectorial operator of angle $0 \leq \phi < (\alpha-1)\frac{\pi}{2}$, where $\varpi < 0$. Suppose further that $f = g + h \in BC(\mathbb{R} \times X, [D(E)])$ with g satisfying the condition (A1) in Section 2.3 of Chapter 2 and $h \in \mathscr{E}(\mathbb{R} \times X, [D(E)])$. If there exists a function $L(\cdot) \in L^1(\mathbb{R}, \mathbb{R}_+)$ such that*

$$\|f(t,x) - f(t,y)\| \leq L(t)\|x - y\|, \text{ for, all } t \in \mathbb{R}, \text{ and } x, y \in X,$$

then Eq. (5.1) has a unique mild solution $u \in PBP_{\omega,k}(\mathbb{R}, [D(E)])$.

Proof. Note that if $t \geq 1$, then

$$\|S_{\alpha,\beta}^{E}(t)\| \leq \frac{M}{|\varpi|}\frac{1}{t^{\alpha-\beta+1}} \leq \frac{M}{|\varpi|},$$

and if $0 \leq t \leq 1$, then

$$\|S_{\alpha,\beta}^{E}(t)\| \leq \frac{M}{1+|\varpi|t^{\alpha}} \leq M.$$

Therefore, $\|S_{\alpha,\beta}^{E}(t)\| \leq N$, where $N = \max\{M, \frac{M}{|\varpi|}\}$.

Define the operator F as (5.6). For each $\Phi \in PBP_{\omega,k}(\mathbb{R}, [D(E)])$, by Corollary 2.3 of Chapter 2 and Theorem 5.4 we have $F\Phi \in PBP_{\omega,k}(\mathbb{R}, [D(E)])$. Thus F is well defined. For $\Phi_1, \Phi_2 \in PBP_{\omega,k}(\mathbb{R}, [D(E)])$ and $t \in \mathbb{R}$ we have

$$\|(F\Phi_1)(t) - (F\Phi_2)(t)\|$$

$$\leq \int_{-\infty}^{t} \|S_{\alpha,\beta}^{E}(t - s)[f(s, \Phi_1(s)) - f(s, \Phi_2(s))]\| ds$$

$$\leq N\|\Phi_1 - \Phi_2\|_\infty \int_{0}^{\infty} L(t - \tau) d\tau$$

$$= N\|\Phi_1 - \Phi_2\|_\infty \int_{-\infty}^{t} L(s) ds.$$

Generally we get

$$\|(F^n\Phi_1)(t) - (F^n\Phi_2)(t)\|$$

$$\leq \|\Phi_1 - \Phi_2\|_\infty \frac{N^n}{(n-1)!} \left(\int_{-\infty}^{t} L(s) \left(\int_{-\infty}^{s} L(\tau) d\tau \right)^{n-1} ds \right)$$

$$\leq \|\Phi_1 - \Phi_2\|_\infty \frac{N^n}{n!} \left(\int_{-\infty}^{t} L(s) ds \right)^n$$

$$\leq \|\Phi_1 - \Phi_2\|_\infty \frac{(\|L\|_{L^1(\mathbb{R},\mathbb{R}_+)} N)^n}{n!},$$

i.e.,

$$\|F^n\Phi_1 - F^n\Phi_2\|_\infty \leq \frac{(\|L\|_{L^1(\mathbb{R},\mathbb{R}_+)} N)^n}{n!} \|\Phi_1 - \Phi_2\|_\infty.$$

Hence, since $\frac{(\|L\|_{L^1(\mathbb{R},\mathbb{R}_+)} N)^n}{n!} < 1$ for n large enough, by the contraction principle F has a unique fixed point $u \in PBP_{\omega,k}(\mathbb{R}, [D(E)])$. $\square$

From Corollary 2.2(2), Corollary 2.4 of Chapter 2 and Theorem 5.4, we can similarly have the following results on weighted pseudo (ω, k)-periodic mild solutions to Eq. (5.1).

Theorem 5.7. *Let $\rho \in \mathcal{U}_{\mathrm{inv}}, 1 \leq \beta < \alpha < 2$ and assume that the pair (A, E) is a ϖ-sectorial operator of angle $0 \leq \phi < (\alpha - 1)\frac{\pi}{2}$, where $\varpi < 0$. Suppose further that $f = g + h \in BC(\mathbb{R} \times X, [D(E)])$ with g satisfying the* *condition (A1) in Section 2.3 of Chapter 2 and* $h \in \mathscr{E}(\mathbb{R} \times X, [D(E)], \rho)$. *If there exists a constant*

$$0 < L < \frac{\alpha}{M}|\varpi|^{\beta/\alpha}\mathbf{B}\left(\frac{\beta}{\alpha}, 1 - \frac{\beta}{\alpha}\right)^{-1}$$

such that (5.5) holds, where M *is the constant given in Theorem 5.3, and* $\mathbf{B}(\cdot, \cdot)$ *denotes the Beta function, then Eq. (5.1) has a unique mild solution* $u \in WPBP_{\omega,k}(\mathbb{R}, [D(E)], \rho)$.

Theorem 5.8. *Let* $\rho \in \mathcal{U}_{\mathrm{inv}}, 1 \le \beta < \alpha < 2$ *and assume that the pair* (A, E) *is a* ϖ*-sectorial operator of angle* $0 \le \phi < (\alpha - 1)\frac{\pi}{2}$, *where* $\varpi < 0$. *Suppose further that* $f = g + h \in BC(\mathbb{R} \times X, [D(E)])$ *with* g *satisfying the condition (A1) in Section 2.3 of Chapter 2 and* $h \in \mathscr{E}(\mathbb{R} \times X, [D(E)], \rho)$. *If there exists a function* $L(\cdot) \in L^1(\mathbb{R}, \mathbb{R}_+)$ *such that the condition (C2) in Corollary 2.4 of Chapter 2 is satisfied, then Eq. (5.1) has a unique mild solution* $u \in WPBP_{\omega,k}(\mathbb{R}, [D(E)], \rho)$.

Finally, we give the existence and uniqueness of S-asymptotically (ω, k)-Bloch periodic mild solutions to Eq. (5.1).

Theorem 5.9. *Let* $1 \le \beta < \alpha < 2$ *and assume that the pair* (A, E) *is a* ϖ*-sectorial operator of angle* $0 \le \phi < (\alpha - 1)\frac{\pi}{2}$, *where* $\varpi < 0$. *Suppose further that* $f \in BC(\mathbb{R} \times X, [D(E)])$ *satisfies* $(T1)$ *in Theorem 2.5 of Chapter 2 and (5.5) with*

$$0 < L < \frac{\alpha}{M}|\varpi|^{\beta/\alpha}\mathbf{B}\left(\frac{\beta}{\alpha}, 1 - \frac{\beta}{\alpha}\right)^{-1},$$

then Eq. (5.1) has a unique mild solution $u \in SABP_{\omega,k}(\mathbb{R}, [D(E)])$.

Proof. The proof is mainly based upon Theorems 2.5 of Chapter 2 and 5.4, and can be conducted analogously to that of Theorem 5.5.   □

Remark 5.2. The existence and uniqueness of (weighted) pseudo S-asymptotically (ω, k)-Bloch periodic mild solutions to Eq. (5.1) can be established similarly to Theorem 5.9 via Theorems 2.8, 2.11 of Chapter 2 and 5.4.

Example 5.1. We consider the following problem:

$$\partial_t^\alpha(m(x)u) = \Delta u + \partial_t^{\alpha-\beta}(m(x)f(t, x)), \text{ in } \mathbb{R} \times \Omega, \tag{5.7}$$

$$u = 0, \text{ in } \mathbb{R} \times \partial\Omega, \tag{5.8}$$

where Ω is a bounded domain in $\mathbb{R}^n$ with a smooth boundary $\partial\Omega$, $m(x) \geq 0$ is a given measurable bounded function on Ω and f is a function on $\mathbb{R} \times \Omega$. We notice that if $m(x) = 0$ over a non empty subset of Ω, then the inverse of the multiplication operator E defined by $Eu(t, x) := m(x)u(t, x)$ is unbounded.

We consider this problem in the space $X = H^{-1}(\Omega)$. By Barbu and Favini (1997, p. 38) there exists a constant $C > 0$ such that

$$\|(Ez - \Delta)^{-1}E\| \leq \frac{C}{1 + |z|},$$

for all $\mathrm{Re}(z) \geq -c(1 + |\mathrm{Im}(z)|)$, where c is a positive constant. Take $\varpi = -c < 0$. Observe that $|z - \varpi| \leq |z| + c \leq K(|z| + 1)$ for a positive constant $K \in \mathbb{N}$. Hence, if $\mathrm{Re}(z) \geq -c(1 + |\mathrm{Im}(z)|)$, then

$$\|(Ez - \Delta)^{-1}E\| \leq \frac{C}{1 + |z|} \leq \frac{KC}{|z - \varpi|}.$$

Observe that $\mathrm{Re}(z) \geq -c(1 + |\mathrm{Im}(z)|)$ represents the right-hand side sector of the complex plane bounded by

$$\gamma_1(t) = -c - te^{i \arctan\left(\frac{1}{c}\right)}$$

and

$$\gamma_2(t) = -c - te^{-i \arctan\left(\frac{1}{c}\right)},$$

for all $t \geq 0$. Therefore, the operator $A := \Delta$ is a $(-c)$-sectorial operator with respect to E of angle $\phi = \arctan\left(\frac{1}{c}\right)$.

If $\frac{2}{\pi}\left(\arctan\left(\frac{1}{c}\right) + \frac{\pi}{2}\right) < \alpha < 2$ and $\beta \geq 1$ such that $1 \leq \beta < \alpha < 2$, then by Theorem 5.2 the pair (A, E) generates an (α, β)-resolvent family $\{S_{\alpha,\beta}^E(t)\}_{t\geq0}$ which satisfies (by Theorem 5.3 and Lemma 5.1) that the function $t \mapsto S_{\alpha,\beta}^E(t)$ belongs to $L^1(\mathbb{R}_+, X)$.

Therefore, if $f \in BC(\mathbb{R} \times X, X)$ satisfies conditions (A1) and (5.5), then the problem (5.7)–(5.8) has a unique mild solution $u \in PBP_{\omega,k}(X)$.

Now, Let $f(t, \phi)(s) := g(t, \phi)(s) + h(t, \phi)(s)$ with

$$g(t, \phi)(s) := \varrho\gamma(t)\phi(s), \quad h(t, \phi)(s) := e^{-t^2}\cos(\phi(s)).$$

Assume that $\varrho > 0$ and $\gamma(t)$ is a bounded T-periodic function, i.e., $\gamma(t+T) = \gamma(t)$, then we have

$$g(t + T, e^{ikT}\phi)(s) = \varrho\gamma(t + T)e^{ikT}\phi(s)$$
$$= \varrho\gamma(t)e^{ikT}\phi(s) = e^{ikT}g(t, \phi)(s).$$

Meanwhile, it is easy to see that $h(t,\phi) \in \mathscr{E}(\mathbb{R} \times X, X)$. Moreover, for $u, v \in X$ there exists a constant $D := D(\Omega)$ (by the Poincaré's inequality) such that

$$\|f(t,u) - f(t,v)\|_X^2 \leq D\|f(t,u) - f(t,v)\|_{L^2(\Omega)}^2$$

$$\leq D\varrho^2(1 + \|\gamma\|_\infty^2)\|u - v\|_X^2.$$

If $L := \sqrt{D}\varrho\sqrt{1 + \|\gamma\|_\infty^2}$, then we can choose $\varrho > 0$ such that

$$L < \frac{\alpha}{KC}|c|^{\beta/\alpha}\mathbf{B}\left(\frac{\beta}{\alpha}, 1 - \frac{\beta}{\alpha}\right)^{-1},$$

and therefore the problem (5.7)–(5.8) admits a unique mild solution $u \in PBP_{\omega,k}(X)$.

Chapter 6

Bloch-Type Periodic Solutions to Fractional Integrodifferential Equations

In this chapter, we mainly consider the existence of generalized Bloch-type periodic solutions to the following semilinear fractional integrodifferential equation:

$$\partial_t^\alpha u(t) = Au(t) + \int_{-\infty}^t a(t-s)Au(s)ds + f(t, u(t)), \quad t \in \mathbb{R}, \qquad (6.1)$$

where ∂_t^α denotes the Weyl fractional derivative of order $\alpha > 0$, the operator A generates an α-resolvent family $\{T_\alpha(t)\}_{t \geq 0}$ on a Banach space X, the kernel function $a(\cdot) \in L^1_{\text{loc}}(\mathbb{R}^+)$.

Equation (6.1) usually arises in the models of viscoelastic materials or memory materials. Bounded solutions to Eq. (6.1) is first studied by Ponce (2013a), in which the existence and uniqueness of (asymptotically) ω-periodic solutions, S-asymptotically ω-periodic solutions, (asymptotically, pseudo) almost periodic and (asymptotically, pseudo) almost automorphic solutions are established when f is a bounded continuous function with certain recurrence. Chang *et al.* (2015a) considered some existence results of weighted pseudo almost automorphic solutions to Eq. (6.1) when f is Stepanov-like weighted pseudo almost automorphic. Alvarez *et al.* (2015) accomplished the existence and uniqueness of weighted pseudo antiperiodic solutions to Eq. (6.1) when f is Stepanov-like weighted pseudo antiperiodic. Oueama-Guengai and N'Guérékata (2018) studied the existence and uniqueness of S-asymptotically ω-periodic and (ω, k)-Bloch periodic solutions to Eq. (6.1) when f is a bounded continuous function satisfying additional conditions. It is noted that some special forms of Eq. (6.1) have also been investigated. For instance, if $\alpha = 1$, then Eq. (6.1) is reduced to the following semilinear integro-differential equation

$$\frac{d}{dt}u(t) = Au(t) + \int_{-\infty}^t a(t-s)Au(s)ds + f(t, u(t)), \quad t \in \mathbb{R}. \qquad (6.2)$$

113

Lizama and N'Guérékata (2010) established sufficient conditions for the existence and uniqueness of bounded solutions, such as (asymptotically) ω-periodic solutions, S-asymptotically ω-periodic solutions, (asymptotically, pseudo) almost periodic and (asymptotically, pseudo) almost automorphic solutions to Eq. (6.2) when f is bounded continuous with certain recurrence. For bounded solutions to Eq. (6.2) with some specific kernels $a(\cdot)$, we refer to Bian *et al.* (2014), Chang *et al.* (2015b), Chang and Ponce (2018), and Lizama and Ponce (2011b).

6.1　Fractional Integrodifferential Equation in the Linear Case

In this section, we give some results on linear equation corresponding to Eq. (6.1), most of which can be found in Ponce (2013a).

Definition 6.1 (Ponce, 2013a). Let A be a closed and linear operator with domain $D(A)$ defined on a Banach space X, and $\alpha > 0$. For a given kernel $a(\cdot) \in L^1_{\text{loc}}(\mathbb{R}^+)$, it is said that A is the generator of an α-resolvent family if there exists $\widetilde{\omega} \geq 0$ and a strongly continuous family $T_\alpha : [0, \infty) \to \mathcal{B}(X)$ such that $\left\{ \frac{\lambda^\alpha}{1+\hat{a}(\lambda)} : \text{Re}\lambda > \widetilde{\omega} \right\} \subseteq \rho(A)$ and for all $x \in X$,

$$(\lambda^\alpha - (1 + \hat{a}(\lambda))A)^{-1} x = \frac{1}{1 + \hat{a}(\lambda)} \left(\frac{\lambda^\alpha}{1 + \hat{a}(\lambda)} - A \right)^{-1} x$$

$$= \int_0^\infty e^{-\lambda t} T_\alpha(t) x \, dt, \qquad \text{Re}\lambda > \widetilde{\omega}.$$

In this case, $\{T_\alpha(t)\}_{t \geq 0}$ is called the α-resolvent family generated by A.

Remark 6.1 (see also Ponce, 2013a, Remark 2.4). Let

$$b(t) = g_\alpha(t) + (g_\alpha * a)(t),$$

where $g_\alpha(t) = \frac{t^{\alpha-1}}{\Gamma(\alpha)}$, $t \geq 0$. Then we have that the α-resolvent family $\{T_\alpha(t)\}_{t \geq 0}$ is a (b, g_α)-regularized family, see Lizama (2000) (see also Definition 1.8 and Lemma 1.17 of Chapter 1). In particular, if $a \equiv 0$, an 1-resolvent family is the same as a C_0-semigroup, whereas that a 2-resolvent family corresponds to the concept of sine family. Thus, if A is the generator of an α-resolvent family $\{T_\alpha(t)\}_{t \geq 0}$, then by Lemmas 1.16 and 1.17 of Chapter 1 (see also Lizama, 2000, Proposition 3.1, Lemma 2.2) we have the following properties:

(i) $T_\alpha(0) = g_\alpha(0)$;
(ii) $T_\alpha(t)x \in D(A)$ and $T_\alpha(t)Ax = AT_\alpha(t)x$ for all $x \in D(A)$ and $t \geq 0$;

(iii) $T_\alpha(t)x = g_\alpha(t)x + \int_0^t b(t-s)AT_\alpha(s)x ds$, for all $x \in D(A)$ and $t \geq 0$;

(iv) $\int_0^t b(t-s)T_\alpha(s)x ds \in D(A)$ and

$$T_\alpha(t)x = g_\alpha(t)x + A \int_0^t b(t-s)T_\alpha(s)x ds,$$

for all $x \in X$ and $t \geq 0$.

We now consider the following linear equation to Eq. (6.1):

$$\partial_t^\alpha u(t) = Au(t) + \int_{-\infty}^t a(t-s)Au(s)ds + f(t), \quad t \in \mathbb{R}. \tag{6.3}$$

Let A generate an α-resolvent family $\{T_\alpha(t)\}_{t \geq 0}$ which is uniformly integrable. For a given $f \in BC(\mathbb{R}, X)$, let $\Phi(t)$ be the function defined by

$$\Phi(t) := \int_{-\infty}^t T_\alpha(t-s)f(s)ds, \quad t \in \mathbb{R}. \tag{6.4}$$

It is obvious that $\|\Phi\|_\infty \leq \|T_\alpha\|_1 \|f\|_\infty$. If $f(t) \in D(A)$ for all $t \in \mathbb{R}$, then $\Phi(t) \in D(A)$ for all $t \in \mathbb{R}$ (see Lemma 1.4 of Chapter 1 or Arendt *et al.*, 2001, Proposition 1.1.7). Take $b(t) = g_\alpha(t) + (g_\alpha * a)(t)$, $n = [\alpha] + 1$ and assume the existence of $\partial_t^\alpha \Phi$. It follows from Remark 6.1 and the Fubini's theorem that for all $t \in \mathbb{R}$

$$\partial_t^\alpha \Phi(t)$$

$$= \frac{d^n}{dt^n} \int_{-\infty}^t g_{n-\alpha}(t-s)\Phi(s)ds$$

$$= \frac{d^n}{dt^n} \int_{-\infty}^t g_{n-\alpha}(t-s) \int_{-\infty}^s T_\alpha(s-r)f(r)drds$$

$$= \frac{d^n}{dt^n} \int_{-\infty}^t g_{n-\alpha}(t-s)$$

$$\times \int_{-\infty}^s \Big[g_\alpha(s-r)f(r) + (b * AT_\alpha)(s-r)f(r) \Big] drds$$

$$= \frac{d^n}{dt^n} \int_{-\infty}^t g_{n-\alpha}(t-s)\partial_t^{-\alpha}f(s)ds + \frac{d^n}{dt^n} \int_{-\infty}^t g_{n-\alpha}(t-s)$$

$$\times \int_{-\infty}^s \int_0^{s-r} b(s-r-v)AT_\alpha(v)f(r)dvdrds$$

$$= f(t) + \frac{d^n}{dt^n} \int_{-\infty}^t g_{n-\alpha}(t-s) \int_{-\infty}^s [g_\alpha(s-w)$$

$$+ (g_\alpha * a)(s - w)] \int_{-\infty}^{w} A T_\alpha(w - r) f(r) dr dw ds$$

$$= f(t) + \frac{d^n}{dt^n} \int_{-\infty}^{t} g_{n-\alpha}(t - s) \int_{-\infty}^{s} g_\alpha(s - w) A\Phi(w) dw ds$$

$$+ \frac{d^n}{dt^n} \int_{-\infty}^{t} g_{n-\alpha}(t - s) \int_{-\infty}^{s} (g_\alpha * a)(s - w) A\Phi(w) dw ds$$

$$= f(t) + A\Phi(t) + \frac{d^n}{dt^n} \int_{-\infty}^{t} g_{n-\alpha}(t - s)$$

$$\times \int_{-\infty}^{s} \int_{-\infty}^{v} g_\alpha(s - v) a(v - w) A\Phi(w) dw dv ds$$

$$= f(t) + A\Phi(t) + \frac{d^n}{dt^n} \int_{-\infty}^{t} g_{n-\alpha}(t - s)$$

$$\times \int_{-\infty}^{s} g_\alpha(s - v)(a \dot{*} A\Phi)(v) dv ds$$

$$= f(t) + A\Phi(t) + (a \dot{*} A\Phi)(t),$$

where $(a \dot{*} A\Phi)(t) := \int_{-\infty}^{t} a(t - s) A\Phi(s) ds$. That is, Φ is a (strict) solution to Eq. (6.3). Generally, $\partial_t^\alpha \Phi$ does not exist or we merely have $f(t) \in X$, in this case, we call that $\Phi(t)$ given by (6.4) is a mild solution to Eq. (6.3).

We recall that the following assumption.

(RNT) Let A generate an α-resolvent family $\{T_\alpha(t)\}_{t \geq 0}$ such that for all $t \geq 0$, $\|T_\alpha(t)\| \leq \varphi_\alpha(t)$, where $\varphi_\alpha(\cdot) \in L^1(\mathbb{R}^+)$.

Theorem 6.1. *Let the condition (RNT) hold and* $f \in \mathcal{N}(X)$*. Then Eq. (6.3) has a unique mild solution* $u \in \mathcal{N}(X)$*.*

Proof. Since the α-resolvent family $\{T_\alpha(t)\}_{t \geq 0}$ generated by A is uniformly integrable, if $f \in \mathcal{N}(X)$, then u given by

$$u(t) := \int_{-\infty}^{t} T_\alpha(t - s) f(s) ds$$

is well defined, and u also belongs to $\mathcal{N}(X)$ via Theorems 2.1, 2.2, 2.7, 2.10, 2.13 of Chapter 2. Thus, u is the unique mild solution to Eq. (6.3). $\quad\square$

In what follows, we consider the following linear fractional differential equation:

$$\partial_t^\alpha u(t) = -\varrho u(t) + f(t), \quad t \in \mathbb{R}, \quad 0 < \alpha < 2. \tag{6.5}$$

Let $e_{\alpha,\beta}(t) = t^{\beta-1}E_{\alpha,\beta}(-\varrho t^\alpha)$, $\varrho \in \mathbb{R}$, where $E_{\alpha,\beta}(z)$ denotes Mittag–Leffler function. For $0 < \alpha \le \beta < 1$, it is known from Gorenflo and Mainardi (1997, p. 268) that

$$e_{\alpha,\beta}(t) = \frac{1}{\pi} \int_0^\infty e^{-rt} K_{\alpha,\beta}(r)dr, \ t \ge 0, \tag{6.6}$$

$$K_{\alpha,\beta}(r) = \frac{[r^\alpha \sin(\beta\pi) + \varrho \sin((\beta - \alpha)\pi)]}{r^{2\alpha} + 2\varrho r^\alpha \cos(\alpha\pi) + \varrho^2} r^{\alpha-\beta}. \tag{6.7}$$

The following result shows that a similar representation to (6.6) of $e_{\alpha,\beta}$ is also valid when $1 < \beta \le \alpha < 2$.

Lemma 6.1 (Ponce, 2013a, Proposition 3.6). *Let* $1 < \beta \le \alpha < 2$, $\varrho \in \mathbb{R}$. *For all* $t \ge 0$ *we have:*

$$e_{\alpha,\beta}(t) = \frac{1}{\pi} \int_0^\infty e^{-rt} K_{\alpha,\beta}(r)dr + \frac{2}{\alpha} \varrho^{(1-\beta)/\alpha} e^{\varrho^{1/\alpha} t \cos(\pi/\alpha)}$$

$$\times \cos\left(\varrho^{1/\alpha} t \sin(\pi/\alpha) + \frac{(1-\beta)\pi}{\alpha}\right), \tag{6.8}$$

where $K_{\alpha,\beta}$ *is defined by* (6.7).

Proof. It follows the same lines as Araya and Lizama (2008) and Gorenflo and Mainardi (1997). From the inversion complex formula for the Laplace transform, we have

$$e_{\alpha,\beta}(t) = \frac{1}{2\pi i} \int_{B_r} e^{\lambda t} \frac{\lambda^{\alpha-\beta}}{\lambda^\alpha + \varrho} d\lambda,$$

where B_r denotes the Bromwich path, i.e., a line $\mathrm{Re}(\lambda) = \sigma \ge \varrho^{1/\alpha}$ and $\mathrm{Im}(\lambda)$ running from $-\infty$ to ∞. As in Gorenflo and Mainardi (1997) we obtain a decomposition of $e_{\alpha,\beta}$ in two parts,

$$e_{\alpha,\beta}(t) = f_{\alpha,\beta}(t) + g_{\alpha,\beta}(t),$$

where by a Titchmarsh's formula (see Mainardi, 2010, p. 225)

$$f_{\alpha,\beta}(t) = -\frac{1}{\pi} \int_0^\infty e^{-rt} \mathrm{Im}\left(\frac{\lambda^{\alpha-\beta}}{\lambda^\alpha + \varrho}\bigg|_{\lambda=re^{i\pi}}\right) dr$$

$$= \frac{1}{\pi} \int_0^\infty e^{-rt} \frac{[r^\alpha \sin(\beta\pi) + \varrho \sin((\beta - \alpha)\pi)]}{r^{2\alpha} + 2\varrho r^\alpha \cos(\alpha\pi) + \varrho^2} r^{\alpha-\beta} dr$$

and

$$g_{\alpha,\beta}(t) = e^{s_0 t} \mathrm{Res}\left(\frac{\lambda^{\alpha-\beta}}{\lambda^\alpha + \varrho}\right)\bigg|_{s_0} + e^{s_1 t} \mathrm{Res}\left(\frac{\lambda^{\alpha-\beta}}{\lambda^\alpha + \varrho}\right)\bigg|_{s_1}$$

$$= \frac{1}{\alpha}(e^{s_0 t} s_0^{1-\beta} + e^{s_1 t} s_1^{1-\beta}),$$

where $s_0 = \varrho^{1/\alpha} e^{i\pi/\alpha}$ and $s_1 = \varrho^{1/\alpha} e^{-i\pi/\alpha}$ are the poles of $\frac{\lambda^{\alpha-\beta}}{\lambda^\alpha + \varrho}$ $(1 < \beta \le \alpha < 2)$. Therefore,

$$g_{\alpha,\beta}(t) = \frac{2}{\alpha} \varrho^{(1-\beta)/\alpha} e^{\varrho^{1/\alpha} t \cos(\pi/\alpha)}$$

$$\times \cos\left(\varrho^{1/\alpha} t \sin(\pi/\alpha) + \frac{(1-\beta)\pi}{\alpha}\right). \tag{6.9}$$

This ends the proof. $\qquad\qquad\qquad\qquad\qquad\qquad\qquad\qquad\qquad\square$

Remark 6.2. We have the following results, see also Ponce (2013a, Remarks 3.7, 3.8).

(i) It is noticed that if $0 < \beta \le \alpha < 1$, then $\frac{\lambda^{\alpha-\beta}}{\lambda^\alpha + \varrho}$ has no poles, and consequently $g_{\alpha,\beta}(t) = 0$, $t \ge 0$. Thus, if $0 < \beta \le \alpha < 1$, then

$$e_{\alpha,\beta}(t) = f_{\alpha,\beta}(t) = \frac{1}{\pi} \int_0^\infty e^{-rt} K_{\alpha,\beta}(r) dr.$$

(ii) Since for $1 < \beta \le \alpha < 2$, $0 = e_{\alpha,\beta}(0) = f_{\alpha,\beta}(0) + g_{\alpha,\beta}(0)$, it follows from (6.8) and (6.9) that

$$\frac{1}{\pi} \int_0^\infty K_{\alpha,\beta}(r) dr = -\frac{2}{\alpha} \varrho^{(1-\beta)/\alpha} \cos\left(\frac{(1-\beta)\pi}{\alpha}\right). \tag{6.10}$$

Lemma 6.2 (Ponce, 2013a, Lemma 3.9). *If $0 < \beta \le \alpha < 1$ and $\varrho > 0$, then $e_{\alpha,\beta} \in L^1(\mathbb{R}_+)$.*

Proof. By Remark 6.2, $g_{\alpha,\beta}(t) = 0$ for all $t \ge 0$. First, we prove the lemma in the case $0 < \beta < \alpha < 1$. An easy computation using complex analysis (see Lang, 1999, p. 199) shows that if $0 < |\gamma| < 1$ and $0 < |\theta| < \pi$ then

$$\int_0^\infty \frac{x^\gamma}{x^2 + 2xa \cos\theta + a^2} dx = a^{\gamma-1} \frac{\pi}{\sin\gamma\pi} \frac{\sin\gamma\theta}{\sin\theta}, \quad a > 0. \tag{6.11}$$

By Fubini's theorem and facts

$$r^{2\alpha} + 2\varrho r^\alpha \cos(\alpha\pi) + \varrho^2 = (r^\alpha \cos(\alpha\pi) + \varrho)^2 + (r^\alpha \sin(\alpha\pi))^2 \ge 0,$$

$$K_{\alpha,\beta}(r) = r K_{\alpha,\beta+1}(r),$$

we get

$$\int_0^\infty |f_{\alpha,\beta}(t)| dt$$

$$\leq \frac{1}{\pi} \int_0^\infty \int_0^\infty e^{-rt} |K_{\alpha,\beta}(r)| dr dt$$

$$= \frac{1}{\pi} \int_0^\infty r^{-1} |K_{\alpha,\beta}(r)| dr$$

$$= \frac{1}{\pi} \int_0^\infty |K_{\alpha,\beta+1}(r)| dr$$

$$= \frac{1}{\pi} \int_0^\infty \frac{|r^\alpha \sin((\beta+1)\pi) + \varrho \sin(((\beta+1) - \alpha)\pi)|}{r^{2\alpha} + 2\varrho r^\alpha \cos(\alpha\pi) + \varrho^2} r^{\alpha-(\beta+1)} dr$$

$$\leq \frac{1}{\pi} \int_0^\infty \frac{r^{2\alpha-(\beta+1)}}{r^{2\alpha} + 2\varrho r^\alpha \cos(\alpha\pi) + \varrho^2} dr$$

$$+ \frac{\varrho}{\pi} \int_0^\infty \frac{r^{\alpha-(\beta+1)}}{r^{2\alpha} + 2\varrho r^\alpha \cos(\alpha\pi) + \varrho^2} dr$$

$$:= I_1 + \varrho I_2.$$

By a substitution of variable $x = r^\alpha$, we obtain by (6.11) that

$$I_1 = \frac{1}{\alpha\pi} \int_0^\infty \frac{x^{1-\beta/\alpha}}{x^2 + 2x\varrho \cos(\alpha\pi) + \varrho^2} dx$$

$$= \frac{1}{\alpha} \varrho^{-\beta/\alpha} \frac{\sin((\alpha-\beta)\pi)}{\sin \alpha\pi \sin \frac{\beta}{\alpha}\pi},$$

and

$$I_2 = \frac{1}{\alpha\pi} \int_0^\infty \frac{x^{-\beta/\alpha}}{x^2 + 2x\varrho \cos(\alpha\pi) + \varrho^2} dx$$

$$= \frac{1}{\alpha} \varrho^{-(1+\beta/\alpha)} \frac{\sin \beta\pi}{\sin \alpha\pi \sin \frac{\beta}{\alpha}\pi},$$

since $-1 < 1 - \beta/\alpha < 1$ and $-1 < -\beta/\alpha < 1$ when $0 < \beta < \alpha < 1$. Hence,

$$\int_0^\infty |f_{\alpha,\beta}(t)| dt \leq \frac{1}{\alpha} \varrho^{-\beta/\alpha} \frac{\sin((\alpha-\beta)\pi)}{\sin \alpha\pi \sin \frac{\beta}{\alpha}\pi}$$

$$+ \frac{1}{\alpha} \varrho^{-\beta/\alpha} \frac{\sin \beta\pi}{\sin \alpha\pi \sin \frac{\beta}{\alpha}\pi} < \infty.$$

For $0 < \beta = \alpha < 1$, we have

$$\int_0^\infty |f_{\alpha,\alpha}(t)| dt \leq \frac{1}{\pi} \int_0^\infty |K_{\alpha,\alpha+1}(r)| dr$$

$$= \frac{\sin(\alpha\pi)}{\pi} \int_0^\infty \frac{r^{\alpha-1}}{r^{2\alpha} + 2\varrho r^\alpha \cos(\alpha\pi) + \varrho^2} dr$$

$$= \frac{\sin(\alpha\pi)}{\alpha\pi} \varrho^{-1} \int_0^\infty \frac{1}{x^2 + 2x \cos(\alpha\pi) + 1} dx$$

$$= \frac{\varrho^{-1}}{\alpha\pi} \left(\frac{\pi}{2} - \arctan(\cot(\alpha\pi)) \right) < \infty.$$

The proof is ended.   □

Lemma 6.3 (Ponce, 2013a, Lemma 3.10). *If* $1 < \beta \le \alpha < 2$ *and* $\varrho > 0$, *then* $e_{\alpha,\beta} \in L^1(\mathbb{R}_+)$.

Proof. The case $\alpha = \beta$ appears in the proof of Araya and Lizama (2008, Corollary 3.7). It is shown the lemma for $1 < \beta < \alpha < 2$ here. From Lemma 6.1, $e_{\alpha,\beta}(t) = f_{\alpha,\beta}(t) + g_{\alpha,\beta}(t)$. Since

$$|g_{\alpha,\beta}(t)| \le \frac{2}{\alpha} \varrho^{(1-\beta)/\alpha} e^{\varrho^{1/\alpha} t \cos(\pi/\alpha)}$$

and $\cos(\pi/\alpha) < 0$ for $1 < \alpha < 2$, we obtain that

$$\int_0^\infty |g_{\alpha,\beta}(t)| dt < \infty.$$

Since for $1 < \beta < \alpha < 2$, $0 < 1 - \beta/\alpha < 1/2$ and $-1 < -\beta/\alpha < -1/2$, we have that $f_{\alpha,\beta} \in L^1(\mathbb{R}_+)$ as in Lemma 6.2 using (6.11).   □

The following result is a consequence of Theorem 6.1, Lemmas 6.2 and 6.3. The case $\alpha = 1$ can be referred to Lizama and N'Guérékata (2010, Corollary 3.6).

Corollary 6.1 (Ponce, 2013a, Corollary 3.11). *Let* $f \in \mathcal{N}(X)$ *and* $\varrho > 0$. *Then, for all* $0 < \alpha < 2$ *Eq.* (6.5) *admits a unique mild solution* u *which belongs to the same space as that of* f *and is given by*

$$u(t) = \int_{-\infty}^t S_{\alpha,\alpha}(t - s) f(s) ds, \quad t \in \mathbb{R},$$

where, for $0 < \alpha < 1$,

$$S_{\alpha,\alpha}(t) = \frac{1}{\pi} \sin \pi\alpha \int_0^\infty e^{-rt} \frac{r^\alpha}{r^{2\alpha} + 2r^\alpha \varrho \cos \pi\alpha + \varrho^2} dr, \quad t \ge 0, \quad (6.12)$$

and for $1 \le \alpha < 2$,

$$S_{\alpha,\alpha}(t) = \frac{1}{\pi} \int_0^\infty e^{-rt} K_{\alpha,\alpha}(r) dr - \frac{2}{\alpha} \varrho^{(1-\alpha)/\alpha} e^{\varrho^{1/\alpha} t \cos(\pi/\alpha)}$$

$$\times \cos \left(\varrho^{1/\alpha} t \sin \left(\frac{\pi}{\alpha} \right) + \frac{\pi}{\alpha} \right), \quad t \ge 0. \quad (6.13)$$

Remark 6.3 (Ponce, 2013a, Remark 3.12). For $1 < \alpha = \beta < 2$, it follows by Araya and Lizama (2008, p. 3700) that

$$\int_0^\infty |e_{\alpha,\alpha}(t)| dt \le \frac{2}{\alpha\varrho} - \frac{1}{\varrho} - \frac{2}{\alpha\varrho} \frac{1}{\cos(\pi/\alpha)} := l(\alpha, \varrho). \quad (6.14)$$

6.2 Fractional Integrodifferential Equation in the Semilinear Case

In this section, we consider the fractional semilinear integrodifferential equation (6.1).

Definition 6.2. Let $\alpha > 0$ and A be the generator of an α-resolvent family $\{T_\alpha(t)\}_{t\geq 0}$. A function $u \in C(\mathbb{R}, X)$ is called a mild solution to Eq. (6.1) if the function $s \mapsto T_\alpha(t-s)f(s,u(s))$ is integrable on $(-\infty, t)$ for each $t \in \mathbb{R}$ and

$$u(t) = \int_{-\infty}^{t} T_\alpha(t-s)f(s,u(s))ds, \quad t \in \mathbb{R}.$$

Theorem 6.2. *Let the condition (RNT) hold. Suppose further that $f = g + h \in BC(\mathbb{R} \times X, X)$ satisfies the Lipschitz condition*

$$\|f(t,x) - f(t,y)\| \leq L\|x - y\|, \quad \text{for all } t \in \mathbb{R} \quad \text{and} \quad x,y \in X, \quad (6.15)$$

where $0 < L < \|\varphi_\alpha\|_{L^1(\mathbb{R}_+)}^{-1}$ and g satisfies the condition (A1) in Section 2.3 of Chapter 2, $h \in \mathscr{E}(\mathbb{R} \times X, X)$. Then Eq. (6.1) has a unique mild solution $u \in PBP_{\omega,k}(\mathbb{R}, X)$.

Proof. Define the operator $\Upsilon : PBP_{\omega,k}(\mathbb{R}, X) \to PBP_{\omega,k}(\mathbb{R}, X)$ by

$$(\Upsilon u)(t) := \int_{-\infty}^{t} T_\alpha(t-s)f(s,u(s))\,ds, \quad t \in \mathbb{R}. \qquad (6.16)$$

For each $u \in PBP_{\omega,k}(\mathbb{R}, X)$, by (1) in Corollary 2.2 of Chapter 2, the function $s \mapsto f(s,u(s))$ belongs to $PBP_{\omega,k}(\mathbb{R}, X)$. It follows from Theorem 6.1 that $\Upsilon u \in PBP_{\omega,k}(\mathbb{R}, X)$, which implies that Υ is well defined.

For $u_1, u_2 \in PBP_{\omega,k}(\mathbb{R}, X)$ and $t \in \mathbb{R}$, we have

$$\|(\Upsilon u_1)(t) - (\Upsilon u_2)(t)\|$$

$$\leq \int_{-\infty}^{t} \|T_\alpha(t-s)[f(s,u_1(s)) - f(s,u_2(s))]\|ds$$

$$\leq \int_{-\infty}^{t} L\|T_\alpha(t-s)\| \cdot \|u_1(s) - u_2(s)\|ds$$

$$\leq L\|u_1 - u_2\|_\infty \int_{-\infty}^{t} \varphi_\alpha(t-s)ds$$

$$\leq L\|u_1 - u_2\|_\infty \|\varphi_\alpha\|_{L^1(\mathbb{R}_+)}.$$

Thus,

$$\|\Upsilon u_1 - \Upsilon u_2\|_\infty \leq L\|u_1 - u_2\|_\infty \|\varphi_\alpha\|_{L^1(\mathbb{R}_+)},$$

which shows that Υ is a contraction. By the Banach fixed point theorem there exists a unique $u \in PBP_{\omega,k}(\mathbb{R}, X)$ such that $\Upsilon u = u$. $\qquad\square$

As a direct consequence of Theorem 6.2 and Corollary 6.1, we have the following corollary.

Corollary 6.2. *Let $\varrho > 0$ and $f = g + h \in BC(\mathbb{R} \times X, X)$ satisfy (6.15), where $0 < L < l(\alpha, \varrho)^{-1}$, and $l(\alpha, \varrho)$ is defined by (6.14), g satisfies the condition (A1) in Section 2.3 of Chapter 2, $h \in \mathscr{E}(\mathbb{R} \times X, X)$. Then, for all $0 < \alpha < 2$ the equation*
$$\partial_t^\alpha u(t) = -\varrho u(t) + f(t, u(t)), \quad t \in \mathbb{R},$$
has a unique mild solution $u \in PBP_{\omega,k}(\mathbb{R}, X)$ which is expressed as
$$u(t) = \int_{-\infty}^t S_{\alpha,\alpha}(t - s)f(s, u(s))ds, \quad t \in \mathbb{R},$$
where $S_{\alpha,\alpha}$ is given by (6.12) if $0 < \alpha < 1$ and by (6.13) if $1 \leq \alpha < 2$.

Recall that a resolvent family $\{S(t)\}_{t \geq 0} \subseteq \mathcal{B}(X)$ is said to be uniformly bounded if there exists a constant $N > 0$ such that $\|S(t)\| \leq N$ for all $t \geq 0$. See for instance $\{S_{\alpha,\beta}(t)\}_{t \geq 0}$ in Theorem 4.6 of Chapter 4.

We investigate some existence and uniqueness of $PBP_{\omega,k}$ mild solutions to Eq. (6.1) under different Lipschitz conditions.

Theorem 6.3. *Let A generate a uniformly bounded α-resolvent family $\{T_\alpha(t)\}_{t \geq 0}$ satisfying the condition (RNT). Let $f = g + h \in BC(\mathbb{R} \times X, X)$ verify the following condition:*
$$\|f(t, x) - f(t, y)\| \leq L_f(t)\|x - y\|, \quad \text{for all } t \in \mathbb{R} \quad \text{and} \quad x, y \in X, \tag{6.17}$$
where $L_f \in L^1(\mathbb{R}, \mathbb{R}_+)$, g satisfies the condition (A1) in Section 2.3 of Chapter 2 and $h \in \mathscr{E}(\mathbb{R} \times X, X)$. Then Eq. (6.1) has a unique mild solution $u \in PBP_{\omega,k}(\mathbb{R}, X)$.

Proof. We define the operator $\Upsilon : PBP_{\omega,k}(\mathbb{R}, X) \to PBP_{\omega,k}(\mathbb{R}, X)$ as (6.16). By Corollary 2.3 of Chapter 2, for each $u \in PBP_{\omega,k}(\mathbb{R}, X)$, the function $s \mapsto f(s, u(s))$ belongs to $PBP_{\omega,k}(\mathbb{R}, X)$. It follows from Theorem 6.1 that $\Upsilon u \in PBP_{\omega,k}(\mathbb{R}, X)$, and thus Υ is well defined. We now show Υ is a contraction.

For $\phi_1, \phi_2 \in PBP_{\omega,k}(\mathbb{R}, X)$ and $t \in \mathbb{R}$ we have
$$\|(\Upsilon\phi_1)(t) - (\Upsilon\phi_2)(t)\|$$
$$\leq \int_{-\infty}^t \|T_\alpha(t - s)[f(s, \phi_1(s)) - f(s, \phi_2(s))]\|ds$$

$$\leq N\|\phi_1 - \phi_2\|_\infty \int_0^\infty L_f(t-\tau)d\tau$$

$$= N\|\phi_1 - \phi_2\|_\infty \int_{-\infty}^t L_f(s)ds.$$

Generally we get

$$\|(\Upsilon^n\phi_1)(t) - (\Upsilon^n\phi_2)(t)\|$$

$$\leq \|\phi_1 - \phi_2\|_\infty \frac{N^n}{(n-1)!} \left(\int_{-\infty}^t L_f(s) \left(\int_{-\infty}^s L_f(\tau)d\tau \right)^{n-1} ds \right)$$

$$\leq \|\phi_1 - \phi_2\|_\infty \frac{N^n}{n!} \left(\int_{-\infty}^t L_f(s)ds \right)^n$$

$$\leq \|\phi_1 - \phi_2\|_\infty \frac{(N\|L_f\|_{L^1(\mathbb{R},\mathbb{R}_+)})^n}{n!},$$

i.e.

$$\|\Upsilon^n\phi_1 - \Upsilon^n\phi_2\|_\infty \leq \frac{(N\|L_f\|_{L^1(\mathbb{R},\mathbb{R}_+)})^n}{n!}\|\phi_1 - \phi_2\|_\infty.$$

Since $\frac{(N\|L\|_{L^1(\mathbb{R},\mathbb{R}_+)})^n}{n!} < 1$ for n sufficiently large, by the contraction principle (see Corollary 1.1 of Chapter 1), Υ has a unique fixed point $u \in PBP_{\omega,k}(\mathbb{R}, X)$. □

Theorem 6.4. *Assume that the α-resolvent family $\{T_\alpha(t)\}_{t\geq 0}$ generated by A is uniformly exponentially stable. Let $f = g + h \in BC(\mathbb{R} \times X, X)$ verify the condition (6.17), where $L_f \in L^p(\mathbb{R}, \mathbb{R}_+)$ ($1 \leq p < +\infty$), g satisfies the condition (A1) in Section 2.3 of Chapter 2 and $h \in \mathscr{E}(\mathbb{R} \times X, X)$. Then Eq. (6.1) has a unique mild solution $u \in PBP_{\omega,k}(\mathbb{R}, X)$.*

Proof. Define the operator $\Upsilon : PBP_{\omega,k}(\mathbb{R}, X) \to PBP_{\omega,k}(\mathbb{R}, X)$ as (6.16). Since $\{T_\alpha(t)\}_{t\geq 0}$ is uniformly exponentially stable, there exist constants $C, \delta > 0$ such that $\|T_\alpha(t)\| \leq Ce^{-\delta t}$ for all $t \geq 0$. Thus the condition (RNT) is satisfied. It follows by Corollary 2.3 of Chapter 2 and Theorem 6.1 that $\Upsilon u \in PBP_{\omega,k}(\mathbb{R}, X)$ for each $u \in PBP_{\omega,k}(\mathbb{R}, X)$. For $p = 1$, it is obvious that $\{T_\alpha(t)\}_{t\geq 0}$ is uniformly bounded and can be completed the proof just as Theorem 6.3.

For $L_f(\cdot) \in L^p(\mathbb{R}, \mathbb{R}_+)$ with $1 < p < \infty$, let $\tau(t) = \int_{-\infty}^t L_f^p(s)ds$. Define an equivalent norm over $PBP_{\omega,k}(\mathbb{R}, X)$ by

$$\|u\|_\tau = \sup_{t\in\mathbb{R}} \left\{ e^{-\theta\tau(t)}\|u\|_\infty \right\}, \quad u \in PBP_{\omega,k}(\mathbb{R}, X),$$

where $\theta > 0$, is a constant sufficiently large. Now for each $u, v \in PBP_{\omega,k}(\mathbb{R}, X)$, we have

$$\|(\Upsilon u)(t) - (\Upsilon v)(t)\|$$

$$\leq \int_{-\infty}^{t} \|T_\alpha(t - s)[f(s, u(s)) - f(s, v(s))]\| ds$$

$$\leq C \int_{-\infty}^{t} e^{-\delta(t-s)} L_f(s) \|u(s) - v(s)\| ds$$

$$\leq C \int_{-\infty}^{t} e^{-\delta(t-s)} L_f(s) e^{\theta \tau(s)} \|u - v\|_\tau ds$$

$$\leq C \left[\int_{-\infty}^{t} e^{\theta p \tau(s)} L_f^p(s) ds \right]^{\frac{1}{p}} \left[\int_{-\infty}^{t} e^{-\delta \frac{p(t-s)}{p-1}} ds \right]^{\frac{p-1}{p}} \|u - v\|_\tau$$

$$\leq C \left(\delta \frac{p}{p-1} \right)^{\frac{1-p}{p}} \left[\int_{-\infty}^{t} e^{\theta p \tau(s)} d\tau(s) \right]^{\frac{1}{p}} \|u - v\|_\tau$$

$$\leq C \left(\delta \frac{p}{p-1} \right)^{\frac{1-p}{p}} (p\theta)^{-\frac{1}{p}} e^{\theta \tau(t)} \|u - v\|_\tau.$$

Consequently,

$$\|\Upsilon u - \Upsilon v\|_\tau \leq C \left(\delta \frac{p}{p-1} \right)^{\frac{1-p}{p}} (p\theta)^{-\frac{1}{p}} \|u - v\|_\tau.$$

We can see that the operator Υ is a contraction for a sufficiently large θ. Thus, there exists a unique $u \in PBP_{\omega,k}(\mathbb{R}, X)$ such that $u = \Upsilon u$ via the Banach fixed point theorem. $\square$

Theorem 6.5. *Let A generate an α-resolvent family $\{T_\alpha(t)\}_{t \geq 0}$ satisfying $\|T_\alpha(t)\| \leq \varphi_\alpha(t)$ for all $t \geq 0$, where $\varphi_\alpha : \mathbb{R}_+ \to \mathbb{R}_+$ is a decreasing function such that*

$$\varphi_0 := \sum_{m=0}^{\infty} \varphi_\alpha(m) < \infty.$$

Assume further that $f = g + h \in BC(\mathbb{R} \times X, X)$ verify the condition (6.17), where $L_f \in L^1(\mathbb{R}, \mathbb{R}_+)$, g satisfies the condition (A1) in Section 2.3 of Chapter 2 and $h \in \mathscr{E}(\mathbb{R} \times X, X)$. Then Eq. (6.1) has a unique mild solution $u \in PBP_{\omega,k}(\mathbb{R}, X)$ whenever

$$\bar{L}\varphi_0 < 1 \text{ with } \bar{L} := \sup_{t \in \mathbb{R}} \int_t^{t+1} L_f(s) ds.$$

Proof. Since φ_α is nonincreasing and $\varphi_0 := \sum_{m=0}^\infty \varphi_\alpha(m) < \infty$, we conclude that $\varphi_\alpha \in L^1(\mathbb{R}_+)$ and thus the condition (RNT) holds. Define the operator $\Upsilon : PBP_{\omega,k}(\mathbb{R}, X) \to PBP_{\omega,k}(\mathbb{R}, X)$ as (6.16). It follows by Corollary 2.3 of Chapter 2 and Theorem 6.1 that $\Upsilon u \in PBP_{\omega,k}(\mathbb{R}, X)$ for each $u \in PBP_{\omega,k}(\mathbb{R}, X)$. Let $u, v \in PBP_{\omega,k}(\mathbb{R}, X)$, we have

$$\|(\Upsilon u)(t) - (\Upsilon v)(t)\|$$

$$= \left\| \int_{-\infty}^t T_\alpha(t-s)[f(s, u(s)) - f(s, v(s))]ds \right\|$$

$$\leq \int_{-\infty}^t L_f(s)\|T_\alpha(t-s)\|\|u(s) - v(s)\|ds$$

$$\leq \left(\sum_{m=0}^\infty \int_{t-(m+1)}^{t-m} L_f(s)\varphi_\alpha(t-s)ds \right) \|u - v\|_\infty$$

$$\leq \left(\sum_{m=0}^\infty \varphi_\alpha(m) \int_{t-(m+1)}^{t-m} L_f(s)ds \right) \|u - v\|_\infty$$

$$\leq \bar{L}\varphi_0\|u - v\|_\infty,$$

i.e., $\|\Upsilon u - \Upsilon v\|_\infty \leq \bar{L}\varphi_0\|u - v\|_\infty$. We can complete the proof by the Banach fixed point theorem whenever $\bar{L}\varphi_0 < 1$. □

Remark 6.4. Let $\rho \in \mathcal{U}_{\text{inv}}$ and take into account Theorem 6.1, Corollaries 2.2(2) and 2.4 of Chapter 2, we can obtain that Eq. (6.1) admits a unique mild solution $u \in WPBP_{\omega,k}(\mathbb{R}, X, \rho)$ such as Theorems 6.2–6.5.

Theorem 6.6. *Let the condition (RNT) hold. Suppose further that $f \in BC(\mathbb{R} \times X, X)$ satisfies the condition (T1) in Theorem 2.5 of Chapter 2 and (6.15) with*

$$0 < L < \|\varphi_\alpha\|_{L^1(\mathbb{R}_+)}^{-1}.$$

Then Eq. (6.1) has a unique mild solution $u \in SABP_{\omega,k}(\mathbb{R}, X)$.

Proof. Define the operator $F : SABP_{\omega,k}(\mathbb{R}, X) \to SABP_{\omega,k}(\mathbb{R}, X)$ by

$$(Fu)(t) := \int_{-\infty}^t T_\alpha(t-s)f(s, u(s))ds, \quad t \in \mathbb{R}.$$

According to Theorems 2.5 of Chapter 2 and 6.1, Fu belongs to $SABP_{\omega,k}(\mathbb{R}, X)$ whenever $u \in SABP_{\omega,k}(\mathbb{R}, X)$. Thus the operator F is well defined. It remains to show that F is a contraction on $SABP_{\omega,k}(\mathbb{R}, X)$, which can be accomplished just as Theorem 6.2. □

As a consequence of Theorem 6.6 and Corollary 6.1, we have the following result.

Corollary 6.3. *Let $\varrho > 0$ and $f \in BC(\mathbb{R} \times X, X)$ satisfy the condition (T1) in Theorem 2.5 of Chapter 2 and (6.15) with $0 < L < l(\alpha, \varrho)^{-1}$, where $l(\alpha, \varrho)$ is defined by (6.14). Then, for all $0 < \alpha < 2$ the equation*

$$\partial_t^\alpha u(t) = -\varrho u(t) + f(t, u(t)), \quad t \in \mathbb{R},$$

has a unique mild solution $u \in SABP_{\omega,k}(\mathbb{R}, X)$ which is expressed as

$$u(t) = \int_{-\infty}^t S_{\alpha,\alpha}(t - s)f(s, u(s))ds, \quad t \in \mathbb{R},$$

where $S_{\alpha,\alpha}$ is given by (6.12) if $0 < \alpha < 1$ and by (6.13) if $1 \leq \alpha < 2$.

Remark 6.5. Based upon Theorems 2.8, 2.11 of Chapter 2, and 6.1, we can show that Eq. (6.1) admits a unique mild solution $u \in PSABP_{\omega,k}(\mathbb{R}, X)$ (or $WPSABP_{\omega,k}(\mathbb{R}, X, \rho)$) similarly to Theorem 6.6.

Finally, we list some concrete examples to show the uniform integrablity of α-resolvent families in the scalar case, which can also be found in Ponce (2013a, Examples 4.17 and 4.18).

Example 6.1. Let $A = -\varrho I$ and $a(t) = \frac{\varrho}{4}\frac{t^{\alpha-1}}{\Gamma(\alpha)}$, with $0 < \alpha < 1$ and $\varrho > 0$. From Eq. (6.3) we have

$$\partial_t^\alpha u(t) = -\varrho u(t) - \frac{\varrho^2}{4}\int_{-\infty}^t \frac{(t-s)^{\alpha-1}}{\Gamma(\alpha)}u(s)ds + g(t), \quad t \in \mathbb{R}. \quad (6.18)$$

It follows from the Laplace transform and Remark 6.1 that A generates an α-resolvent family $\{T_\alpha(t)\}_{t \geq 0}$ satisfying

$$\widehat{T_\alpha}(\lambda) = \frac{\lambda^\alpha}{(\lambda^\alpha + \varrho/2)^2} = \frac{\lambda^{\alpha-\alpha/2}}{(\lambda^\alpha + \varrho/2)} \cdot \frac{\lambda^{\alpha-\alpha/2}}{(\lambda^\alpha + \varrho/2)}.$$

Hence, $T_\alpha(t) = (r * r)(t)$, where

$$r(t) = t^{\frac{\alpha}{2}-1}E_{\alpha,\alpha/2}\left(-\frac{\varrho}{2}t^\alpha\right).$$

By Lemma 6.2, $T_\alpha \in L^1(\mathbb{R}_+)$. Moreover, if $g \in SABP_{\omega,k}(\mathbb{R}, X)$, then Eq. (6.18) has a unique mild solution $u \in SABP_{\omega,k}(\mathbb{R}, X)$ via Theorem 6.1. Suppose further that $f \in BC(\mathbb{R} \times X, X)$ satisfies the condition (T1) in Theorem 2.5 of Chapter 2 and (6.15) with $L < \|T_\alpha\|^{-1}$, then there exists a unique mild solution $u \in SABP_{\omega,k}(\mathbb{R}, X)$ to the following semilinear equation:

$$\partial_t^\alpha u(t) = -\varrho u(t) - \frac{\varrho^2}{4}\int_{-\infty}^t \frac{(t-s)^{\alpha-1}}{\Gamma(\alpha)}u(s)ds + f(t, u(t)), \quad t \in \mathbb{R}.$$

by Theorem 6.2.

Example 6.2. Take $\alpha = 1/2$, $A = -\varrho I$ and $a(t) = \gamma e^{-\beta t}$, with $\beta, \varrho > 0$ and $\gamma \in \mathbb{R}$ in Eq. (6.1), that is

$$\partial_t^{\frac{1}{2}} u(t) = -\varrho u(t) - \gamma \varrho \int_{-\infty}^{t} e^{-\beta(t-s)} u(s) ds$$

$$+ f(t, u(t)), \quad t \in \mathbb{R}. \tag{6.19}$$

From the Laplace transform and Remark 6.1, we deduce that A generates an $1/2$−resolvent family $\{T_{1/2}(t)\}_{t \geq 0}$ such that

$$\widehat{T}_{1/2}(\lambda) = \frac{\lambda + \beta}{\lambda^{3/2} + \lambda \varrho + \lambda^{1/2} \beta + \varrho(\beta + \gamma)}$$

$$= \frac{\lambda + \beta}{(\lambda^{1/2} - r_1)(\lambda^{1/2} - r_2)(\lambda^{1/2} - r_3)},$$

where r_1, r_2, r_3 are the roots (real or complex) of

$$z^3 + \varrho z^2 + \beta z + \varrho(\beta + \gamma) = 0. \tag{6.20}$$

Note that

$$\widehat{T}_{1/2}(\lambda) = \frac{\lambda^{1/2 - 1/6}}{(\lambda^{1/2} - r_1)} \cdot \frac{\lambda^{1/2 - 1/6}}{(\lambda^{1/2} - r_2)} \cdot \frac{\lambda^{1/2 - 1/6}}{(\lambda^{1/2} - r_3)}$$

$$+ \frac{\beta}{(\lambda^{1/2} - r_1)(\lambda^{1/2} - r_2)(\lambda^{1/2} - r_3)},$$

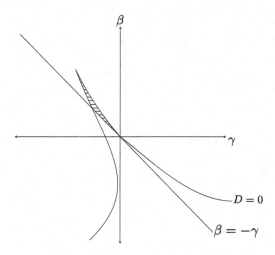

Fig. 6.1   Integrability region.

and therefore,

$$T_{1/2}(t) = (R_1 * R_2 * R_3)(t) + \beta(S_1 * S_2 * S_3)(t),$$

where

$$\begin{cases} R_i(t) = t^{1/6-1} E_{1/2,1/6}\big(r_i t^{1/2}\big), \\ S_i(t) = t^{1/2-1} E_{1/2,1/2}\big(r_i t^{1/2}\big), \quad i = 1,2,3. \end{cases}$$

If $r_i < 0$ $(i = 1,2,3)$, then $T_{1/2} \in L^1(\mathbb{R}_+)$ by Lemma 6.2.

We recall that the discriminant of Eq. (6.20) is given by

$$D := \varrho^2(\beta + \gamma)[18\beta - 4\varrho^2 - 27(\beta + \gamma)] + \beta^2(\varrho^2 - 4\beta),$$

and if $D \geq 0$, then the equation (6.20) has three real roots. It is known from Ponce (2013a, Example 4.18) that if

$$D \geq 0 \quad \text{and} \quad \gamma > -\beta, \tag{6.21}$$

then all roots r_1, r_2, r_3 of (6.20) are negative. Thus it is concluded that if ϱ, β and γ verify the condition (6.21), and $f \in BC(\mathbb{R} \times X, X)$ satisfies the condition (T1) in Theorem 2.5 of Chapter 2 and (6.15) with $L < \|T_{1/2}\|^{-1}$, then Eq. (6.20) has a unique mild solution $u \in SABP_{\omega,k}(\mathbb{R}, X)$ via Theorem 6.2.

A description of the area in the plane where we can choose β and γ in order to have uniform integrability of $T_{1/2}(t)$ for $\varrho > 0$, is shown in the hatched area in Fig. 6.1.

Chapter 7

Asymptotically Bloch-Type Periodic Solutions to Damped Evolution Equations

In this chapter, we mainly consider the following damped evolution equation:

$$\begin{cases} \alpha u'''(t) + u''(t) = \beta A u(t) + \gamma A u'(t) + f(t, u(t)), & t \geq 0, \\ u(0) = 0, u'(0) = y, u''(0) = z, \end{cases} \tag{7.1}$$

where $\alpha, \beta, \gamma \in (0, \infty), \alpha\beta < \gamma$, y, z belong to a Banach space X, the operator A generates an (α, β, γ)-regularized family $\mathcal{R}(t)$, and $f : [0, \infty) \times X \to X$ is a suitable bounded continuous function.

Equation (7.1) can be applied to deal with vibrations of elastic structure having internal material damping and external forcing, see for instance Bose and Gorain (1998a, 1998b) and Gorain (2006). Studies on asymptotic behavior of solutions to Eq. (7.1) was initiated by Andrade and Lizama (2011), in which they established some sufficient conditions for existence of asymptotically almost periodic mild solutions to Eq. (7.1) via an (α, β, γ)-regularized family. Andrade $et\ al.$ (2016) subsequently studied the existence of pseudo S-asymptotically ω-periodic mild solutions to Eq. (7.1), and showed some concrete applications to flexible structures. The aforementioned works (Andrade and Lizama, 2011; Andrade $et\ al.$, 2016) show that for each asymptotically almost periodic/pseudo S-asymptotically ω-periodic input disturbance f, the output mild solution u to Eq. (7.1) still remains asymptotically almost periodic/pseudo S-asymptotically ω-periodic. The aim of this chapter is to investigate the existence of pseudo S-asymptotically Bloch-type periodic mild solutions to Eq. (7.1) by introducing pseudo S-asymptotically Bloch-type periodic function on the nonnegative real axis (see Chen $et\ al.$, 2022 for details). Our results can be considered as a useful supplement to works in Andrade and Lizama (2011), Andrade $et\ al.$ (2016), and Chang and Wei (2021b).

Nonlocal Cauchy problems originate from some physical phenomena which possess better effects than the classical Cauchy problems, see for instance (Byszewski and Lakshmikantham, 1991, Section 3). The significance of nonlocal Cauchy problems to various differential equations can be found in Benedetti *et al.* (2020), Brindle and N'Guérékata (2020), Burlică *et al.* (2016), Cao and Huang (2018), Chang and Li (2006), Chang *et al.* (2009), Chen *et al.* (2018), Ding *et al.* (2008), Fan and Li (2010), Fu and Zhang (2013), Hernández and O'Regan (2018), Lizama and Rueda (2019) and references cited therein. Particularly, Cao and Huang (2018) studied a semilinear evolution equation with nonlocal initial conditions and presented some new results on the existence of asymptotically ω-periodic mild solutions. Brindle and N'Guérékata (2020) considered a semilinear fractional differential equation with nonlocal initial conditions and established some new results for the existence of S-asymptotically ω-periodic mild solutions. Inspired by works in Brindle and N'Guérékata (2020) and Cao and Huang (2018), we continue to investigate the existence of pseudo S-asymptotically Bloch-type periodic solutions to the following nonlocal Cauchy problem:

$$\begin{cases} \alpha u'''(t) + u''(t) = \beta Au(t) + \gamma Au'(t) + f(t, u(t)), & t \geq 0, \\ u(0) = 0, u'(0) = 0, u''(0) = g(u), \end{cases} \tag{7.2}$$

where $g : BC(\mathbb{R}_+, X) \to X$ will be specified in the later.

7.1 Some Basic Results

We first give some properties on the space $PSABP_{\omega,k}(\mathbb{R}_+, X)$ which can be analogously attained from Chang and Wei (2021b).

Definition 7.1. A function $f \in BC(\mathbb{R}_+, X)$ is said to be pseudo S-asymptotically Bloch-type periodic (or pseudo S-asymptotically (ω, k)-Bloch periodic) if for given $\omega \in \mathbb{R}_+$, $k \in \mathbb{R}$,

$$\lim_{T \to \infty} \frac{1}{T} \int_0^T \left\| f(t + \omega) - e^{ik\omega} f(t) \right\| dt = 0$$

holds for each $t \in \mathbb{R}_+$. We denote the space of all such functions by $PSABP_{\omega,k}(\mathbb{R}_+, X)$.

Remark 7.1. Let $k\omega = \pi$ in Definition 7.1, we can obtain the notion of pseudo S-asymptotically ω-antiperiodic function on $\mathbb{R}_+$.

Lemma 7.1. *Let* $f_1, f_2, f \in PSABP_{\omega,k}(\mathbb{R}_+, X)$. *Then the following results hold:*

(I) $f_1 + f_2 \in PSABP_{\omega,k}(\mathbb{R}_+, X)$, *and* $cf \in PSABP_{\omega,k}(\mathbb{R}_+, X)$ *for each* $c \in \mathbb{R}$.

(II) $f_a := f(t + a) \in PSABP_{\omega,k}(\mathbb{R}_+, X)$ *for each* $a \in \mathbb{R}$ *such that* $t + a \in \mathbb{R}_+$.

(III) *The space* $PSABP_{\omega,k}(\mathbb{R}_+, X)$ *is a Banach space with the sup-norm.*

Lemma 7.2. *Let* $f \in BC(\mathbb{R}_+ \times X, X)$ *satisfy the following conditions:*

(E1) *For all* $(t, x) \in \mathbb{R}_+ \times X$, $f(t + \omega, x) = e^{ik\omega} f\left(t, e^{-ik\omega} x\right)$.

(E2) *There exists a constant* $L > 0$ *such that for all* $x, y \in X$ *and* $t \in \mathbb{R}_+$,

$$\|f(t, x) - f(t, y)\| \leq L\|x - y\|.$$

Then for each $\phi \in PSABP_{\omega,k}(\mathbb{R}_+, X)$, *we have*

$$f(\cdot, \phi(\cdot)) \in PSABP_{\omega,k}(\mathbb{R}_+, X).$$

Lemma 7.3. *Let* $f \in BC(\mathbb{R}_+ \times X, X)$ *satisfy the condition (E1) and the following condition:*

(E3) *For any* $\varepsilon > 0$ *and any bounded subset* $\mathcal{Q} \subseteq X$, *there exist constants* $T_{\varepsilon,\mathcal{Q}} \geq 0$ *and* $\delta_{\varepsilon,\mathcal{Q}} > 0$ *such that*

$$\|f(t, x) - f(t, y)\| \leq \varepsilon$$

for all $x, y \in \mathcal{Q}$ *with* $\|x - y\| \leq \delta_{\varepsilon,\mathcal{Q}}$ *and* $t \geq T_{\varepsilon,\mathcal{Q}}$.

Then for each $\phi \in PSABP_{\omega,k}(\mathbb{R}_+, X)$, *we have*

$$f(\cdot, \phi(\cdot)) \in PSABP_{\omega,k}(\mathbb{R}_+, X).$$

Remark 7.2. The condition (E1) can be weakened by the following condition:

(E1′) For all $(t, x) \in \mathbb{R}_+ \times X$,

$$\lim_{T \to \infty} \frac{1}{T} \int_0^T \left\| f(t + \omega, x) - e^{ik\omega} f\left(t, e^{-ik\omega} x\right) \right\| dt = 0$$

uniformly on any bounded set of X.

Remark 7.3. If $k\omega = \pi$, we can get some composition theorems for pseudo S-asymptotically ω-antiperiodic functions from Lemmas 7.2–7.3 and Remark 7.2. On the other hand, if $k\omega = 2\pi$ and conditions (E1′), (E3) hold, then the conclusion of Lemma 7.3 is just a slightly modification of Henríquez *et al.* (2008a, Lemma 4.1) for S-asymptotically ω-periodic functions.

Lemma 7.4. *Let $\{\mathcal{S}(t)\}_{t\geq 0} \subseteq \mathcal{B}(X)$ be uniformly exponentially stable, i.e., there exist positive constants M, δ such that $\|\mathcal{S}(t)\| \leq Me^{-\delta t}$ for $t \in \mathbb{R}_+$. If $f \in PSABP_{\omega,k}(\mathbb{R}_+, X)$, then*

$$u(t) := \int_0^t \mathcal{S}(t-s)f(s)ds \in PSABP_{\omega,k}(\mathbb{R}_+, X).$$

Proof. Since $\{\mathcal{S}(t)\}_{t\geq 0} \subseteq \mathcal{B}(X)$ is uniformly exponentially stable, we have $\int_0^\infty e^{-\delta t}dt < \infty$ and $\|u(t)\| \leq \frac{M}{\delta}\|f\|_\infty$. For any $t \in \mathbb{R}_+$, it follows from $f \in PSABP_{\omega,k}(\mathbb{R}_+, X)$ that:

$$\lim_{T\to\infty} \frac{1}{T} \int_0^T \left\| f(t+\omega) - e^{ik\omega} f(t) \right\| dt = 0. \tag{7.3}$$

On the other hand, by the Fubini theorem, we have

$$\frac{1}{T} \int_0^T \left\| u(t+\omega) - e^{i\omega k} u(t) \right\| dt$$

$$= \frac{1}{T} \int_0^T \left\| \int_0^{t+\omega} \mathcal{S}(t+\omega-s)f(s)ds - \int_0^t \mathcal{S}(t-s)e^{i\omega k} f(s)ds \right\| dt$$

$$\leq \frac{1}{T} \int_0^T \left\| \int_0^t \mathcal{S}(t-s) \left[f(s+\omega) - e^{i\omega k} f(s) \right] ds \right\| dt$$

$$+ \frac{1}{T} \int_0^T \left\| \int_0^\omega \mathcal{S}(t-s+\omega)f(s)ds \right\| dt$$

$$\leq \frac{M}{T} \int_0^T \left[\int_0^t e^{-\delta(t-s)} \left\| f(s+\omega) - e^{i\omega k} f(s) \right\| ds \right] dt$$

$$+ \frac{M}{T} \int_0^T \left[\int_0^\omega e^{-\delta(t+\omega-s)} \left\| f(s) \right\| ds \right] dt$$

$$\leq \frac{M}{T} \int_0^T \left[\int_0^t e^{-\delta s} \left\| f(t-s+\omega) - e^{i\omega k} f(t-s) \right\| ds \right] dt$$

$$+ \frac{M}{T} \int_0^T \left[\int_t^{t+\omega} e^{-\delta s} \left\| f(t+\omega-s) \right\| ds \right] dt$$

$$\leq M \int_0^\infty e^{-\delta s} \left[\frac{1}{T} \int_0^T \left\| f(t-s+\omega) - e^{i\omega k} f(t-s) \right\| dt \right] ds$$

$$+ \frac{M\|f\|_\infty \omega}{T\delta}.$$

It follows from (7.3), Lemma 7.1 (II) and the Lebesgue dominated convergence theorem that:

$$\lim_{T \to \infty} \frac{1}{T} \int_0^T \left\| u(t+\omega) - e^{i\omega k} u(t) \right\| dt = 0,$$

i.e., $u \in PSABP(\mathbb{R}_+, X)$. $\qquad\qquad\qquad\qquad\qquad\qquad\qquad\qquad\square$

In the following, we recall some fundamental results on regularized families of bounded operator which are established in Andrade and Lizama (2011) and Andrade *et al.* (2016).

Definition 7.2. Let A be a closed and linear operator with domain $D(A)$ defined on a Banach space X. The operator A is said to be the generator of an (α, β, γ)-regularized family $\{\mathcal{R}(t)\}_{t \geq 0} \subseteq \mathcal{B}(X)$ if it verifies the following conditions:

(R1) $\mathcal{R}(t)$ is strongly continuous on $[0, \infty)$ and $\mathcal{R}(0) = 0$;
(R2) $\mathcal{R}(t)D(A) \subset D(A)$ and $A\mathcal{R}(t)x = \mathcal{R}(t)Ax$ for all $x \in D(A), t \geq 0$;
(R3) $\mathcal{R}(t)x = k(t)x + \int_0^t a(t-s)\mathcal{R}(s)Ax \, ds$ for all $x \in D(A), t \geq 0$, where

$$k(t) = -\alpha + t + \alpha e^{-t/\alpha},$$
$$a(t) = -(\alpha\beta - \gamma) + \beta t + (\alpha\beta - \gamma)e^{-t/\alpha}, \quad t \geq 0.$$

Remark 7.4 (see also Andrade and Lizama, 2011, Remark 2.2). In the more general context of (a, k)-regularized families (Lizama, 2000), it can be shown that the operator A generates an (α, β, γ)-regularized family if and only if there exist $\delta > 0$ and a strongly continuous function $\mathcal{R} : \mathbb{R}_+ \to \mathcal{B}(X)$ such that $\left\{ \frac{\lambda^2 + \alpha\lambda^3}{\beta + \gamma\lambda} : \mathrm{Re}\lambda > \delta \right\} \subset \rho(A)$ and

$$H(\lambda)x := \frac{1}{\beta + \gamma\lambda} \left(\frac{\lambda^2 + \alpha\lambda^3}{\beta + \gamma\lambda} I - A \right)^{-1} x$$
$$= \int_0^\infty e^{-\lambda t} \mathcal{R}(t)x \, dt, \quad \mathrm{Re}\lambda > \delta, \quad x \in X.$$

From the uniqueness of the Laplace transform, it is noted that an (α, β, γ)-regularized family corresponds to an (a, k)-regularized family in Lizama (2000). In fact, for all $\mathrm{Re}\lambda > \delta$ we have

$$\hat{a}(\lambda) = \frac{\beta + \gamma\lambda}{\lambda^2 + \alpha\lambda^3}, \quad \hat{k}(\lambda) = \frac{1}{\lambda^2 + \alpha\lambda^3}.$$

The following result is a direct consequence of Lemmas 1.16–1.17 (see also Lizama, 2000, Proposition 3.1, Lemma 2.2).

Lemma 7.5. *Let* $\{\mathcal{R}(t)\}_{t\geq 0}$ *be an* (α, β, γ)-*regularized family on* X *with the generator* A. *Then the following results hold:*

(a) *For all* $x \in D(A)$, *we have* $\mathcal{R}(\cdot)x \in C^2(\mathbb{R}_+, X)$.

(b) *Let* $x \in X$ *and* $t \geq 0$. *Then* $\int_0^t a(t-s)\mathcal{R}(s)x ds \in D(A)$ *and*

$$\mathcal{R}(t)x = k(t)x + A\int_0^t a(t-s)\mathcal{R}(s)x ds.$$

The linear equation corresponding to Eq. (7.1) is given as

$$\begin{cases} \alpha u'''(t) + u''(t) = \beta A u(t) + \gamma A u'(t) + f(t), & t \geq 0, \\ u(0) = 0, u'(0) = y, u''(0) = z. \end{cases} \tag{7.4}$$

For a strong solution to Eq. (7.4), it implies that a function u belongs to $C(\mathbb{R}_+, D(A)) \cap C^3(\mathbb{R}_+, X)$ with $u' \in C(\mathbb{R}_+, D(A))$ and verifying Eq. (7.4).

The following result is from Andrade *et al.* (2016, Proposition 2.2), which gives a complete description of the solutions to Eq. (7.4) via the (α, β, γ)-regularized family $\{\mathcal{R}(t)\}_{t\geq 0}$. We can also refer to Andrade and Lizama (2011, Proposition 3.1) for a more general existence result (i.e., $u(0) = x$, $x \in D(A^3)$) and its detailed proof.

Lemma 7.6. *Let* $\{\mathcal{R}(t)\}_{t\geq 0}$ *be an* (α, β, γ)-*regularized family on* X *generated by the operator* A. *If* $y, z \in D(A^2)$ *and* $f \in L^1_{\text{loc}}(\mathbb{R}_+, D(A^2))$, *then* $u(t)$ *given by*

$$u(t) = \mathcal{R}(t)y + \alpha \mathcal{R}'(t)y + \alpha \mathcal{R}(t)z + \int_0^t \mathcal{R}(t-s)f(s)ds, \quad t \geq 0,$$

is a solution to Eq. (7.4).

7.2 Asymptotically Bloch-Type Periodic Solutions

This section is mainly concerned with some existence results of pseudo S-asymptotically (ω, k)-periodic mild solutions to Eqs. (7.1) and (7.2). We recall the following assumption suggested in Andrade and Lizama (2011) and Andrade *et al.* (2016).

(H) There exist constants $M, \delta > 0$ such that the (α, β, γ)-regularized family $\{\mathcal{R}(t)\}_{t\geq 0}$ generated by the operator A satisfy

$$\|\mathcal{R}(t)\| + \|\mathcal{R}'(t)\| \leq Me^{-\delta t}, \quad t \geq 0.$$

The following lemma is similar to Andrade *et al.* (2016, Theorem 3.1).

Lemma 7.7. *Let the (α, β, γ)-regularized family $\{\mathcal{R}(t)\}_{t \geq 0}$ satisfy the assumption (H). If $f \in PSABP_{\omega,k}(\mathbb{R}_+, X)$ is such that $f(t) \in D(A^2)$ for all $t \in \mathbb{R}_+$, then Eq. (7.4) with $y, z \in D(A^2)$ has a unique strong solution $u \in PSABP_{\omega,k}(\mathbb{R}_+, X)$.*

Proof. Since $f \in PSABP_{\omega,k}(\mathbb{R}_+, X)$ with $f \in D(A^2)$ and $y, z \in D(A^2)$, it follows by Lemma 7.6 that the solution to Eq. (7.4) is given by

$$u(t) = \mathcal{R}(t)y + \alpha \mathcal{R}'(t)y + \alpha \mathcal{R}(t)z + \int_0^t \mathcal{R}(t - s)f(s)ds.$$

By Lemma 2.7, we have $t \mapsto \displaystyle\int_0^t \mathcal{R}(t - s)f(s)ds \in PSABP_{\omega,k}(\mathbb{R}_+, X)$. Moreover, it is deduced that

$$\frac{1}{T}\int_0^T \left\| \mathcal{R}'(t + \omega)y - e^{i\omega k}\mathcal{R}'(t)y \right\| dt \leq \frac{2M}{T}\|y\|\int_0^T e^{-\delta t}dt \leq \frac{2M}{\delta T}\|y\|,$$

which implies that

$$\lim_{T \to \infty} \frac{1}{T}\int_0^T \left\| \mathcal{R}'(t + \omega)y - e^{i\omega k}\mathcal{R}'(t)y \right\| dt = 0.$$

Analogously we can obtain that $\mathcal{R}(\cdot)y \in PSABP_{\omega,k}(\mathbb{R}_+, X)$. Thus the strong solution $u \in PSABP_{\omega,k}(\mathbb{R}_+, X)$. The uniqueness is easy to prove by the assumption (H). □

Remark 7.5. Let $\omega k = \pi$ in Lemma 7.7, we can get the existence and uniqueness of pseudo S-asymptotically ω-antiperiodic strong solutions to Eq. (7.4).

Definition 7.3. A continuous function $u : \mathbb{R}_+ \to X$ is said to be the mild solution to Eq. (7.1) if for all $y, z \in X$, $t \geq 0$, it satisfies the following integral equation:

$$u(t) = \mathcal{R}(t)y + \alpha \mathcal{R}'(t)y + \alpha \mathcal{R}(t)z + \int_0^t \mathcal{R}(t - s)f(s, u(s))ds.$$

Theorem 7.1. *Let $\{\mathcal{R}(t)\}_{t \geq 0}$ satisfy the assumption (H). If $f \in BC(\mathbb{R}_+ \times X, X)$ satisfies (E1)–(E2) in Lemma 7.2. Then Eq. (7.1) admits a unique mild solution $u \in PSABP_{\omega,k}(\mathbb{R}_+, X)$ whenever $\frac{LM}{\delta} < 1$.*

Proof. Let $\mathcal{F} : PSABP_{\omega,k}(\mathbb{R}_+, X) \to PSABP_{\omega,k}(\mathbb{R}_+, X)$ be defined by

$$(\mathcal{F}u)(t) := \mathcal{R}(t)y + \alpha\mathcal{R}'(t)y + \alpha\mathcal{R}(t)z + \int_0^t \mathcal{R}(t-s)f(s, u(s))ds. \quad (7.5)$$

From the proof of Lemma 7.7, it follows that $\mathcal{R}(\cdot)y, \alpha\mathcal{R}(\cdot)z, \alpha\mathcal{R}'(\cdot)y \in PSABP_{\omega,k}(\mathbb{R}_+, X)$. For each $u \in PSABP(\mathbb{R}_+, X)$, first by Lemma 7.2, the function $s \mapsto f(s, u(s)) \in PSABP_{\omega,k}(\mathbb{R}_+, X)$, and then by Lemma 7.4, the term

$$\int_0^t \mathcal{R}(t-s)f(s, u(s))ds \in PSABP_{\omega,k}(\mathbb{R}_+, X).$$

Thus, $\mathcal{F}u \in PSABP_{\omega,k}(\mathbb{R}_+, X)$ and $\mathcal{F}$ is well defined. It remains to show that $\mathcal{F}$ admits a unique fixed point in $PSABP_{\omega,k}(\mathbb{R}_+, X)$.

For $u_1, u_2 \in PSABP_{\omega,k}(\mathbb{R}_+, X)$ and $t \in \mathbb{R}_+$, we have

$$\|(\mathcal{F}u_1)(t) - (\mathcal{F}u_2)(t)\|$$

$$\leq \int_0^t \|\mathcal{R}(t-s)[f(s, u_1(s)) - f(s, u_2(s))]\|ds$$

$$\leq ML \int_0^t e^{-\delta(t-s)}\|u_1(s) - u_2(s)\|ds$$

$$\leq ML\|u_1 - u_2\|_\infty \int_0^t e^{-\delta s}ds$$

$$\leq \frac{ML}{\delta}\|u_1 - u_2\|_\infty.$$

Therefore,

$$\|\mathcal{F}u_1 - \mathcal{F}u_2\|_\infty \leq \frac{ML}{\delta}\|u_1 - u_2\|_\infty,$$

which shows that $\mathcal{F}$ is a contraction from the assumption $\frac{ML}{\delta} < 1$. Thus there exists a unique $u \in PSABP_{\omega,k}(\mathbb{R}, X)$ such that $\mathcal{F}u = u$ via the Banach fixed point theorem. $\qquad\square$

Next we investigate the existence of pseudo S-asymptotically (ω, k)-Bloch periodic mild solutions to Eq. (7.1) when f is not necessarily Lipschitz with its state variable but has the following growth condition. We shall adopt the space given by Eq. (1.4) in Chapter 1.

(C) There exists a continuous nondecreasing function $W : \mathbb{R}_+ \to \mathbb{R}_+$ such that

$$\|f(t, x)\| \leq W(\|x\|) \quad \text{for all } t \in \mathbb{R}_+ \quad \text{and} \quad x \in X.$$

Theorem 7.2. *Let the assumption (H) hold and f satisfies conditions (C) and (E1), (E3) in Lemma 7.3. Suppose further that:*

(E4) *For each $\xi > 0$, $\lim_{t\to\infty} \dfrac{1}{h(t)} \displaystyle\int_0^t e^{-\delta(t-s)} W(\xi h(s)) ds = 0$.*

(E5) *For each $\epsilon > 0$, there exists $\zeta > 0$ such that for every $u, v \in C_h(X)$, $\|u - v\|_h \le \zeta$ implies*

$$\sup_{t\ge 0} M \int_0^t e^{-\delta(t-s)} \|f(s, u(s)) - f(s, v(s))\| ds \le \epsilon.$$

(E6) *For each $c \ge 0, r > 0$, the set $\{\mathcal{R}(c-t)f(t,x) : t \in [0,c], x \in X, \|x\| \le r\}$ is relatively compact in X; and for each $\xi \in [0,c]$, $[\mathcal{R}(c+s-\xi) - \mathcal{R}(c-\xi)]f(\xi, x) \to 0$ as $s \to 0$ uniformly for $\|x\| \le r$.*

(E7) *$\lim_{\tau\to\infty} \dfrac{\tau}{\phi(\tau)} > 1$, where*

$$\phi(\tau) := \left\| \|\mathcal{R}(\cdot)y\| + \|\alpha\mathcal{R}'(\cdot)y\| + \|\alpha\mathcal{R}(\cdot)z\| \right.$$
$$\left. + M \int_0^{\cdot} e^{-\delta(\cdot-s)} W(\tau h(s)) ds \right\|_h.$$

Then Eq. (7.1) admits at least one mild solution $u \in PSABP_{\omega,k}(\mathbb{R}, X)$.

Proof. Define the operator $\mathcal{F}$ on $C_h(X)$ as (7.5). For each $u \in C_h(X)$, it follows that:

$$\|(\mathcal{F}u)(t)\| \le (1+\alpha)M\|y\| + \alpha M\|z\| + M \int_0^t e^{-\delta(t-s)} W(\|u\|_h h(s)) ds.$$

Thus, the condition (E4) implies that $\mathcal{F} : C_h(X) \to C_h(X)$. For the sake of convenience, we complete the proof by the following steps.

Step 1. It is shown that $\mathcal{F}$ is continuous. For $\epsilon > 0$, let ζ satisfy the condition (E5). For $u, v \in C_h(X)$ with $\|u - v\|_h \le \zeta$, we have

$$\|(\mathcal{F}u)(t) - (\mathcal{F}v)(t)\| \le M \int_0^t e^{-\delta(t-s)} \|f(s, u(s)) - f(s, v(s))\| ds \le \epsilon.$$

Thus,

$$\|\mathcal{F}u - \mathcal{F}v\|_h = \sup_{t\ge 0} \frac{\|(\mathcal{F}u)(t) - (\mathcal{F}v)(t)\|}{h(t)} \le \epsilon,$$

which implies that $\mathcal{F}$ is continuous.

Step 2. It is shown that $\mathcal{F}$ is completely continuous. Let $B_r(Z)$ be a closed ball with center at 0 and radius r in the space Z. Define $V = \mathcal{F}(B_r(C_h(X)))$

by $v = \mathcal{F}(u)$ for $u \in B_r(C_h(X))$ and $V_b = V|_{[0,b]}$ for a given $b > 0$. For each $t \in [0,b]$, we have

$$v(t) = \mathcal{R}(t)y + \alpha\mathcal{R}'(t)y + \alpha\mathcal{R}(t)z + \int_0^t \mathcal{R}(t-s)f(s,u(s))ds$$

$$\in \mathcal{R}(t)y + \alpha\mathcal{R}'(t)y + \alpha\mathcal{R}(t)z + t\overline{co}(\mathcal{N}),$$

where $co(\mathcal{N})$ denotes the convex hull of $\mathcal{N} := \{\mathcal{R}(t-s)f(s,x) : s \in [0,t], \|x\| \leq h(t)r\}$. Thus, $V_b(t) \subseteq \mathcal{R}(t)y + \alpha\mathcal{R}'(t)y + \alpha\mathcal{R}(t)z + t\overline{co}(\mathcal{N})$, is relatively compact in X by the first arguments in (E6). Note that

$$v(t+s) - v(t)$$

$$= (\mathcal{R}(t+s) - \mathcal{R}(t))y + \alpha(\mathcal{R}'(t+s) - \mathcal{R}'(t))y$$

$$+ \alpha(\mathcal{R}(t+s) - \mathcal{R}(t))z$$

$$+ \int_0^t [\mathcal{R}(t+s-\eta) - \mathcal{R}(t-\eta)]f(\eta,u(\eta))d\eta$$

$$+ \int_t^{t+s} \mathcal{R}(t+s-\eta)f(\eta,u(\eta))d\eta.$$

Since $\mathcal{R}(\cdot), \mathcal{R}'(\cdot)$ are strong continuous, for each $\epsilon > 0$, we can choose a suitable constant $\zeta > 0$ such that for $s \leq \zeta$ sufficiently small

$$\|(\mathcal{R}(t+s) - \mathcal{R}(t))y\| \leq \frac{\epsilon}{5},$$

$$\|\alpha(\mathcal{R}'(t+s) - \mathcal{R}'(t))y\| \leq \frac{\epsilon}{5},$$

$$\|\alpha(\mathcal{R}(t+s) - \mathcal{R}(t))z\| \leq \frac{\epsilon}{5},$$

$$\left\|\int_t^{t+s} \mathcal{R}(t+s-\eta)f(\eta,u(\eta))d\eta\right\|$$

$$\leq M \int_t^{t+s} e^{-\delta(t+s-\eta)}W(rh(\eta))d\eta \leq \frac{\epsilon}{5}.$$

Meanwhile, by the second arguments in the condition (E6), we have

$$\int_0^t \|[\mathcal{R}(t+s-\eta) - \mathcal{R}(t-\eta)]f(\eta,u(\eta))d\eta\| \leq \frac{\epsilon}{5}.$$

From above estimates, we can conclude that the set V_b is equicontinuous. On the other hand, it follows from the condition (E4) that

$$\frac{\|v(t)\|}{h(t)} \leq \frac{(\alpha+1)M\|y\| + \alpha M\|z\|}{h(t)}$$

$$+ \frac{M}{h(t)} \int_0^t e^{-\delta(t-s)}W(rh(s))ds \to 0, \quad \text{as } t \to \infty,$$

and this convergence is independent of $u \in B_r(C_h(X))$. Thus we can infer that V is a relatively compact set in $C_h(X)$ by Lemma 1.19 of Chapter 1.

Step 3. It is shown that the set $\mathcal{Q} := \left\{ u^\lambda : u^\lambda = \lambda \mathcal{F}(u^\lambda), \ \lambda \in (0,1) \right\}$ is bounded. Assume that $u^\lambda(\cdot)$ is a solution to equation $u^\lambda = \lambda \mathcal{F}(u^\lambda)$ for some $\lambda \in (0,1)$, we have the estimate

$$\left\| u^\lambda(t) \right\| \leq \phi \left(\left\| u^\lambda(t) \right\|_h \right) h(t).$$

Thus, we get

$$\frac{\left\| u^\lambda(t) \right\|_h}{\phi \left(\left\| u^\lambda(t) \right\|_h \right)} \leq 1,$$

and combining with the condition (E7), we can show the assertion.

Step 4. It is shown that $\mathcal{F}$ has a fixed point u which belongs to $\overline{PSABP_{\omega,k}(\mathbb{R}_+, X)}$. It follows by Lemma 7.3 and the proof of Theorem 7.1 that

$$\mathcal{F}(PSABP_{\omega,k}(\mathbb{R}_+, X)) \subseteq PSABP_{\omega,k}(\mathbb{R}_+, X).$$

Thus, we can consider $\mathcal{F} : \overline{PSABP_{\omega,k}(\mathbb{R}_+, X)} \to \overline{PSABP_{\omega,k}(\mathbb{R}_+, X)}$ and $\mathcal{F}$ is completely continuous from Steps 1–3. Taking into account that $\mathcal{Q}$ is bounded, we can show the assertion by the Leray–Schauder alternative theorem (see Lemma 1.22 of Chapter 1).

Step 5. It is shown that the fixed point $u \in PSABP_{\omega,k}(\mathbb{R}_+, X)$. Let $\{u^n\}_n \in PSABP_{\omega,k}(\mathbb{R}_+, X)$ converge to u by the norm of $C_h(X)$, we have

$$\|\mathcal{F}u^n - u\|_\infty = \|\mathcal{F}u^n - \mathcal{F}u\|_\infty$$

$$\leq \sup_{t \geq 0} \frac{M}{h(t)} \int_0^t e^{-\delta(t-s)} \|f(s, u^n(s)) - f(s, u(s))\| ds.$$

It follows from the condition (E5) that $\mathcal{F}u^n \to u$, as $n \to \infty$ uniformly in $\mathbb{R}_+$. Since $\mathcal{F}u^n \in PSABP_{\omega,k}(\mathbb{R}_+, X)$, we have $u \in PSABP_{\omega,k}(\mathbb{R}_+, X)$, i.e., Eq. (7.1) admits one mild solution $u \in PSABP_{\omega,k}(\mathbb{R}_+, X)$. $\square$

Remark 7.6 (Andrade *et al.*, 2016, Remark 3.10). The following conditions can be used to guarantee that the condition (E6) holds:

(a) The operator $\mathcal{R}(t)$ is compact for $t > 0$ and $\mathcal{R}(\cdot)$ is uniformly continuous on $(0, \infty)$.

(b) For each $a \geq 0$ and $r > 0$, the set $\{f(t, x) : 0 \leq t \leq a, x \in X, \|x\| \leq r\}$ is relatively compact in X.

In order to consider the nonlocal Cauchy problem (7.2), we further assume that:

(E8) There exist constants $L_g > 0, C_g > 0$ with $\alpha M L_g < 1$ such that for all $u, v \in BC(\mathbb{R}_+, X)$

$$\|g(u) - g(v)\| \le L_g \|u - v\|_\infty, \quad \|g(u)\| \le C_g \|u\|_\infty.$$

Definition 7.4. A continuous function $u : \mathbb{R}_+ \to X$ is said to be the mild solution to Eq. (7.2) if it satisfies the following integral equation:

$$u(t) = \alpha \mathcal{R}(t) g(u) + \int_0^t \mathcal{R}(t - s) f(s, u(s)) ds, \quad t \ge 0.$$

Theorem 7.3. *Let assumptions (H) and (E8) hold. If $f \in BC(\mathbb{R}_+ \times X, X)$ satisfies conditions (E1)–(E2) in Lemma 7.2, then Eq. (7.2) admits a unique mild solution $u \in PSABP_{\omega,k}(\mathbb{R}_+, X)$ whenever*

$$M\left(\alpha L_g + \frac{L}{\delta}\right) < 1.$$

Proof. Let $\Theta : PSABP_{\omega,k}(\mathbb{R}_+, X) \to PSABP_{\omega,k}(\mathbb{R}_+, X)$ be defined by

$$(\Theta u)(t) := \alpha \mathcal{R}(t) g(u) + \int_0^t \mathcal{R}(t - s) f(s, u(s)) ds. \qquad (7.6)$$

It follows from conditions (H) and (E8) that

$$\frac{1}{T} \int_0^T \left\| \alpha \mathcal{R}(t + \omega) g(u) - e^{ikw} \alpha \mathcal{R}(t) g(u) \right\| dt \le \frac{2M\alpha}{T\delta} C_g \|u\|_\infty.$$

Thus, we have

$$\lim_{T \to \infty} \frac{1}{T} \int_0^T \left\| \alpha \mathcal{R}(t + \omega) g(u) - e^{ikw} \alpha \mathcal{R}(t) g(u) \right\| dt \to 0,$$

which implies that $\alpha \mathcal{R}(\cdot) g(u) \in PSABP_{\omega,k}(\mathbb{R}_+, X)$. Furthermore, for each $u \in PSABP_{\omega,k}(\mathbb{R}_+, X)$, by Lemmas 7.2 and 7.4, we can infer that

$$\int_0^t \mathcal{R}(t - s) f(s, u(s)) ds \in PSABP_{\omega,k}(\mathbb{R}_+, X).$$

Hence $\Theta u \in PSABP_{\omega,k}(\mathbb{R}_+, X)$ and Θ is well defined. Meanwhile, for each $u_1, u_2 \in PSABP_{\omega,k}(\mathbb{R}_+, X)$ and $t \in \mathbb{R}_+$, we have

$$\|(\Theta u_1)(t) - (\Theta u_2)(t)\| \le \alpha \|\mathcal{R}(t) g(u_1) - \mathcal{R}(t) g(u_2)\|$$

$$+ \int_0^t \|\mathcal{R}(t - s)[f(s, u_1(s)) - f(s, u_2(s))]\| ds$$

$$\leq M\alpha L_g\|u_1 - u_2\|_\infty + ML\int_0^t e^{-\delta(t-s)}\|u_1(s) - u_2(s)\|ds$$

$$\leq M\alpha L_g\|u_1 - u_2\|_\infty + ML\|u_1 - u_2\|_\infty \int_0^t e^{-\delta s}ds$$

$$\leq M\alpha L_g\|u_1 - u_2\|_\infty + \frac{ML}{\delta}\|u_1 - u_2\|_\infty.$$

Therefore,

$$\|\Theta u_1 - \Theta u_2\|_\infty \leq M\left[\alpha L_g + \frac{L}{\delta}\right]\|u_1 - u_2\|_\infty,$$

which implies that Θ is a contraction since $M\left[\alpha L_g + \frac{L}{\delta}\right] < 1$. It follows by the Banach fixed point theorem that there exists a unique $u \in PSABP_{\omega,k}(\mathbb{R}, X)$ such that $\Theta u = u$. $\qquad\square$

In order to obtain the existence of pseudo S-asymptotically (ω, k)-Bloch periodic mild solutions to Eq. (7.2) when f is not necessarily Lipschitz with its state variable, we replace the condition (E7) by the following hypothesis:

(E7$'$) $\lim_{\tau \to \infty} \dfrac{\tau}{\psi(\tau)} > 1$, where

$$\psi(\tau) := \alpha MC_g\tau + M\left\|\int_0^\cdot e^{-\delta(\cdot - s)}W(\tau h(s))ds\right\|_h.$$

Theorem 7.4. *Assume that conditions (H), (C), (E1), (E3)–(E6), (E7$'$) and (E8) are all satisfied. Then Eq. (7.2) has at least one mild solution $u \in PSABP_{\omega,k}(\mathbb{R}, X)$.*

Proof. Decompose the operator $\Theta = \Theta_1 + \Theta_2$ on $C_h(X)$ as

$$(\Theta_1 u)(t) = \alpha\mathcal{R}(t)g(u), \quad (\Theta_2 u)(t) = \int_0^t \mathcal{R}(t - s)f(s, u(s))ds.$$

For each $u \in C_h(X)$, it follows from conditions (E4) and (E8) that Θ is well defined. It is easily shown by conditions (E4), (E7$'$) and (E8) that there exists a positive constant τ satisfying $\Theta(B_\tau(C_h(X))) \subseteq B_\tau(C_h(X))$. The remainder of the proof can be given in the following steps.

Step I. Θ_1 is contractive on $B_\tau(C_h(X))$. For $u_1, u_2 \in B_\tau(C_h(X))$ and $t \in \mathbb{R}_+$, we have

$$\frac{\|(\Theta_1 u_1)(t) - (\Theta_1 u_2)(t)\|}{h(t)} \leq \frac{M\alpha L_g\|u_1 - u_2\|_\infty}{h(t)} \leq M\alpha L_g\|u_1 - u_2\|_h,$$

i.e., $\|\Theta_1 u_1 - \Theta_1 u_2\|_h \leq M\alpha L_g \|u_1 - u_2\|_h$. We can achieve the assertion via the condition (E8).

Step II. Analogously conducting as Steps 1–2 in the proof of Theorem 7.2, we can show that Θ_2 is completely continuous.

Step III. It is shown that Θ has a fixed point

$$u \in \overline{PSABP_{\omega,k}(\mathbb{R}_+, X) \bigcap B_\tau(C_h(X))}^h .$$

It follows by Lemma 7.3 and the proof of Theorem 7.3 that

$$\Theta(PSABP_{\omega,k}(\mathbb{R}_+, X)) \subseteq PSABP_{\omega,k}(\mathbb{R}_+, X).$$

Together with $\Theta(B_\tau(C_h(X))) \subseteq B_\tau(C_h(X))$, we infer that

$$\Theta(PSABP_{\omega,k}(\mathbb{R}_+, X) \bigcap B_\tau(C_h(X)))$$

$$\subseteq PSABP_{\omega,k}(\mathbb{R}_+, X) \bigcap B_\tau(C_h(X)).$$

Moreover,

$$\Theta\left(\overline{PSABP(\mathbb{R}_+, X) \bigcap B_\tau(C_h(X))}^h\right)$$

$$\subseteq \overline{\Theta(PSABP(\mathbb{R}_+, X) \bigcap B_\tau(C_h(X)))}^h$$

$$\subseteq \overline{PSABP_{\omega,k}(\mathbb{R}_+, X) \bigcap B_\tau(C_h(X))}^h .$$

By the Krasnoselskii fixed point theorem (see Lemma 1.23), we conclude that Θ admits a fixed point $u \in \overline{PSABP_{\omega,k}(\mathbb{R}_+, X) \bigcap B_\tau(C_h(X))}^h$.

Step IV. It is shown that $u \in PSABP_{\omega,k}(\mathbb{R}_+, X) \bigcap B_\tau(C_h(X))$. Let $\{u_n\}$ be a sequence in $PSABP_{\omega,k}(\mathbb{R}_+, X) \bigcap B_\tau(C_h(X))$ which converges to u. Taking into account conditions (E5), (E8) and conducting analogously as Step 5 in the proof of Theorem 7.2, we can show that $\Theta u^n \to u$, as $n \to \infty$ uniformly in $\mathbb{R}_+$. $\qquad\square$

Remark 7.7. According to the proof of Theorem 7.4, we have the following arguments:

(1) We can consider more general nonlocal conditions $u(0) = 0, u'(0) = \varpi(u)$, $u''(0) = g(u)$ by similarly imposing additional conditions on ϖ.
(2) We can get some existence results for pseudo S-asymptotically ω-antiperiodic mild solutions to reference equations from Theorems 7.1–7.4 by taking $k\omega = \pi$.

Example 7.1. Let Ω be a bounded domain in $\mathbb{R}^n$ with smooth boundary $\partial\Omega$. Define $X := L^2(\Omega)$ and suppose $0 < \lambda < \nu$. Consider the linear equation

$$u''(t) + \lambda u'''(t) = c^2(\Delta u + \nu \Delta u') + f(t). \tag{7.7}$$

Take $\alpha := \lambda, \beta := c^2, \gamma := c^2\nu$. It is known from [Andrade and Lizama (2011), Example 4.8] that Dirichlet–Laplacian operator $A := \Delta$ can generate an (α, β, γ)-regularized family $\mathcal{R}(t)$ on X, and there exist constants $M > 0, \delta > 0$ such that $\|\mathcal{R}'(t)\| + \|\mathcal{R}(t)\| \leq Me^{-\delta t}, t \in \mathbb{R}_+$. Hence it is deduced by Lemma 7.7 that for each $f \in PSABP_{\omega,k}(\mathbb{R}_+, X)$ with $f(t) \in D(\Delta^2), t \in \mathbb{R}_+$, Eq. (7.7) with initial conditions $u(0) = 0, u'(0) = y \in D(\Delta^2), u''(0) = z \in D(\Delta^2)$ has a unique strong solution $u \in PSABP_{\omega,k}(\mathbb{R}_+, X)$.

Let $f(t, v)(s) := \varrho a(t)v(s)$ for all $v \in X, s, t \in \mathbb{R}_+$ and $\varrho > 0$. Assume that $a(t)$ is a bounded continuous ω-periodic function, then we have

$$f(t + \omega, v)(s) = \varrho a(t + \omega)e^{ik\omega}e^{-ik\omega}v(s) = e^{ik\omega}\varrho a(t)e^{-ik\omega}v(s)$$
$$= e^{ik\omega}f(t, e^{-ik\omega}v)(s).$$

Meanwhile, for $u, v \in X$, we have

$$\|f(t, u) - f(t, v)\|_X^2 \leq \varrho^2 \|a\|_\infty^2 \|u - v\|_X^2.$$

If $L := \varrho\|a\|_\infty$, then we can select a suitable $\varrho > 0$ satisfying $L < \dfrac{\delta}{M}$. By Theorem 7.1, the nonlinear equation

$$u''(t) + \lambda u'''(t) = c^2(\Delta u + \nu \Delta u') + f(t, u(t)), \tag{7.8}$$

with initial conditions $u(0) = 0, u'(0) = y \in X, u''(0) = z \in X$ admits a unique mild solution $u \in PSABP_{\omega,k}(\mathbb{R}_+, X)$. Furthermore, if we take $g(x)(t) := \Sigma_{i=1}^m \frac{\tau}{m}x(t_i)$ for $x \in BC(\mathbb{R}_+, X)$, where $\tau > 0$ and $m \in \mathbb{N}$, then we can choose suitable $\tau, \varrho > 0$ such that $M\left(\alpha\tau + \frac{L}{\delta}\right) < 1$. Thus, Theorem 7.3 implies that Eq. (7.8) with nonlocal initial conditions $u(0) = 0, u'(0) = 0, u''(0) = g(u)$ has a unique mild solution $u \in PSABP_{\omega,k}(\mathbb{R}_+, X)$. On the other hand, for above given f and g, we can take $W(t) := \varrho\|a\|_\infty t$ and make conditions (E7), (E7'), (E8) hold by seeking suitable $\tau, \varrho > 0$. Thus the condition (E6) and Remark 7.6 can ensure that Eq. (7.7) (or (7.8)) with aforementioned (nonlocal) initial conditions has one mild solution $u \in PSABP_{\omega,k}(\mathbb{R}_+, X)$ via Theorem 7.2 (or Theorem 7.4).

Chapter 8

Asymptotically Bloch-Type Periodic Solutions to Partial Integrodifferential Equations

This chapter is mainly concerned with the existence of pseudo S-asymptotically Bloch-type periodic mild solutions to the following partial integrodifferential equation:

$$\begin{cases} u'(t) = Au(t) + \displaystyle\int_0^t B(t-s)u(s)ds + f(t,u(t)), & t \geq 0, \\ u(0) = u_0, \end{cases} \tag{8.1}$$

where $A : D(A) \subset X \to X$, $B(t) : D(B(t)) \subset X \to X$ are linear, closed and densely defined operators on a Banach space X; $D(A) \subset D(B(t))$ for each $t \geq 0$ and $f \in BC(\mathbb{R}_+ \times X, X)$.

The existence and qualitative properties of solutions to Eq. (8.1) have been investigated in several works. For instance, the existence of asymptotically almost periodic and almost periodic solutions to Eq. (8.1) was studied in Hernández and Dos Santos (2006); the existence of asymptotically almost automorphic solutions to Eq. (8.1) with a nonlocal initial condition was established in Ding *et al.* (2008); the existence of square mean asymptotically almost automorphic mild solutions to Eq. (8.1) driven by a Brownian motion was considered in Zhao *et al.* (2013b); the existence and uniqueness of pseudo S-asymptotically ω-periodic mild solutions in the Stepanov sense to Eq. (8.1) was studied in Xia (2015a); and the existence of S-asymptotically ω-periodic mild solutions to Eq. (8.1) with a nonlocal initial condition was set up in Brindle and N'Guérékata (2019c). As an application, we continue to show some existence results of pseudo S-asymptotically Bloch-type periodic solutions to Eq. (8.1) in this chapter via the technique developed in Ding *et al.* (2008).

8.1   Some Basic Results

We first assume that the abstract integrodifferential Cauchy problem

$$
\begin{cases}
u'(t) = Au(t) + \displaystyle\int_0^t B(t-s)u(s)ds, & t \geq 0, \\
u(0) = u_0,
\end{cases}
\tag{8.2}
$$

has associated a resolvent operator family $\{\mathcal{R}(t)\}_{t\geq 0}$ on X.

Definition 8.1 (Hernández and Dos Santos, 2006). A one parameter family $\{\mathcal{R}(t)\}_{t\geq 0}$ of bounded linear operators from X to X is said to be a strongly continuous resolvent operator for Eq. (8.2) if the following conditions are verified:

(1) $\mathcal{R}(0) = I$ (the identity operator), and the function $\mathcal{R}(t)x$ is continuous on $[0, \infty)$ for each $x \in X$.
(2) $\mathcal{R}(t)D(A) \subset D(A)$ for all $t \geq 0$, and for $x \in D(A)$, $A\mathcal{R}(t)x$ is continuous on $[0, \infty)$ and $\mathcal{R}(t)x$ is continuously differentiable on $[0, \infty)$.
(3) For each $x \in D(A)$, the following resolvent equations are satisfied:

$$
\begin{cases}
\mathcal{R}'(t)x = A\mathcal{R}(t)x + \displaystyle\int_0^t B(t-s)\mathcal{R}(s)xds, & t \geq 0, \\
\mathcal{R}'(t)x = \mathcal{R}(t)Ax + \displaystyle\int_0^t B(t-s)\mathcal{R}(s)xds, & t \geq 0.
\end{cases}
$$

In what follows, we always assume that the following condition holds:

(HR) There exist $M, \delta > 0$ such that $\|\mathcal{R}(t)\| \leq Me^{-\delta t}$ for all $t \geq 0$, i.e., $\{\mathcal{R}(t)\}_{t\geq 0}$ is uniformly exponentially stable.

Definition 8.2 (Ding et al., 2008; Hernández and Dos Santos, 2006). A function $u \in C(\mathbb{R}_+, X)$ is said to be a mild solution to Eq. (8.1) if it verifies the following equation:

$$
u(t) = \mathcal{R}(t)u(0) + \int_0^t \mathcal{R}(t-s)f(s, u(s))ds, \quad t \geq 0.
$$

Let $PSABP_{\omega,k}(\mathbb{R}_+, X)$ be the space defined in Definition 7.1, of Chapter 7. According to Lemma 7.4 of Chapter 7, we have the following result.

Lemma 8.1. *Let the condition (HR) hold. If f belongs to the space $PSABP_{\omega,k}(\mathbb{R}_+, X)$, then the function u defined by*

$$
u(t) := \int_0^t \mathcal{R}(t-s)f(s)ds
$$

also belongs to $PSABP_{\omega,k}(\mathbb{R}_+, X)$.

8.2 Asymptotically Bloch-Type Periodic Solutions

In this section, we investigate the existence of generalized Bloch-type mild solutions to Eq. (8.1). We list the following conditions:

(Cf1) $f \in BC(\mathbb{R}_+ \times X, X)$ and for all $(t, x) \in \mathbb{R}_+ \times X$,

$$f(t + \omega, x) = e^{ik\omega} f\left(t, e^{-ik\omega} x\right).$$

(Cf2) There exists a continuous and nondecreasing function L_f : $[0, +\infty) \to [0, +\infty)$ such that for each $r \geq 0$ and all $x, y \in X$ with $\|x\| \leq r, \|y\| \leq r$,

$$\|f(t, x) - f(t, y)\| \leq L_f(r)\|x - y\|, \quad t \in \mathbb{R}_+.$$

(Cf3) $\sup\limits_{r>0} \left[\dfrac{\delta r}{M} - rL_f(r)\right] > \delta\|u_0\| + \sup\limits_{t \geq 0} \|f(t, 0)\|.$

Theorem 8.1. *Assume that conditions (HR) and (Cf1)–(Cf3) are satisfied. Then Eq. (8.1) has a mild solution $u \in PSABP_{\omega,k}(\mathbb{R}_+, X)$.*

Proof. We define $F : PSABP_{\omega,k}(\mathbb{R}_+, X) \to PSABP_{\omega,k}(\mathbb{R}_+, X)$ by

$$(Fu)(t) = \mathcal{R}(t)u_0 + \int_0^t \mathcal{R}(t - s)f(s, u(s))ds$$
$$:= F_1(t) + F_2(t), \quad t \geq 0, \tag{8.3}$$

where $F_1(t) = \mathcal{R}(t)u_0$ and $F_2(t) = \int_0^t \mathcal{R}(t - s)f(s, u(s))ds$.

It follows from the condition (HR) that

$$\frac{1}{T} \int_0^T \left\| \mathcal{R}(t + \omega)u_0 - e^{i\omega k}\mathcal{R}(t)u_0 \right\| dt \leq \frac{2M}{\delta T}\|u_0\|.$$

Thus, the function $F_1(t) \in PSABP_{\omega,k}(\mathbb{R}_+, X)$.

On the other hand, for a given $u \in PSABP_{\omega,k}(\mathbb{R}_+, X)$, which is bounded, we can choose a bounded subset Q of X such that $u(t) \in Q$ for all $t \geq 0$. It follows from the condition (Cf2) that $f(t, u)$ is uniformly continuous on the bounded subset Q uniformly for $t \in \mathbb{R}_+$. Then, Lemma 7.3 of Chapter 7 yields that $f(\cdot, u(\cdot)) \in PSABP_{\omega,k}(\mathbb{R}_+, X)$. It follows from Lemma 8.1 that $F_2(t) \in PSABP_{\omega,k}(\mathbb{R}_+, X)$. Thus, F is well defined for each $u \in PSABP_{\omega,k}(\mathbb{R}_+, X)$.

By the condition (Cf3), there exists a constant $r > 0$ such that

$$\frac{\delta r}{M} - rL_f(r) > \delta\|u_0\| + \sup\limits_{s \in \mathbb{R}_+} \|f(s, 0)\|. \tag{8.4}$$

Let $\mathbb{D} = \{u \in PSABP_{\omega,k}(\mathbb{R}_+, X) : \|u\|_\infty \leq r\}$. Then $\mathbb{D}$ is a closed subspace of $PSABP_{\omega,k}(\mathbb{R}_+, X)$. We claim that $F(\mathbb{D}) \subseteq \mathbb{D}$. In fact, for any given $u \in \mathbb{D}$ and $t \in \mathbb{R}_+$, we have

$$\|(Fu)(t)\|$$

$$\leq M\|u_0\| + M \int_0^t e^{-\delta(t-s)}[\|f(s,0)\| + \|f(s,u(s)) - f(s,0)\|]ds$$

$$\leq M\|u_0\| + \frac{M}{\delta}\Big[\sup_{s\in\mathbb{R}_+} \|f(s,0)\| + rL_f(r)\Big],$$

which from (8.4) implies that $\|Fu\|_\infty \leq r$ and thus $F(\mathbb{D}) \subseteq \mathbb{D}$.

Now we show that F is a contraction on $\mathbb{D}$. It is known from (8.4) that

$$\frac{\delta r}{M} - rL_f(r) > 0.$$

Thus, we have

$$\frac{M}{\delta}L_f(r) < 1. \qquad (8.5)$$

For each $u, v \in \mathbb{D}$ and all $t \geq 0$, we have

$$\|(Fu)(t) - (Fv)(t)\| \leq \int_0^t \|\mathcal{R}(t-s)[f(s,u(s)) - f(s,v(s))]\|ds$$

$$\leq M \int_0^t e^{-\delta(t-s)}L_f(r)\|u(s) - v(s)\|ds$$

$$\leq \frac{M}{\delta}L_f(r)\|u - v\|_\infty.$$

Hence

$$\|Fu - Fv\|_\infty \leq \frac{M}{\delta}L_f(r)\|u - v\|_\infty.$$

It is deduced by (8.5) that F is a contraction on $\mathbb{D}$. Thus F admits a unique fixed point in $\mathbb{D}$, which is naturally a pseudo S-asymptotically (ω, k)-periodic mild solution to Eq. (8.1). $\qquad\square$

Corollary 8.1. *Suppose that conditions (HR) and (Cf1)–(Cf2) hold. If $L_f(\cdot) \equiv L$ and $0 < L < \frac{\delta}{M}$, then Eq. (8.1) admits a unique mild solution $u \in PSABP_{\omega,k}(\mathbb{R}_+, X)$.*

Proof. Noting that

$$\delta\|u_0\| + \sup_{s\in\mathbb{R}_+} \|f(s,0)\| < \infty,$$

since $0 < L < \frac{\delta}{M}$, there exists a constant $r_0 > 0$ such that for all $r \geq r_0$, we have

$$\frac{\delta r}{M} - rL > \delta \|u_0\| + \sup_{s \in \mathbb{R}_+} \|f(s, 0)\|.$$

It follows from the proof of Theorem 8.1 that there exists a unique mild solution $u \in PSABP_{\omega,k}(\mathbb{R}_+, X)$ to Eq. (8.1). $\qquad\square$

Next, we consider the following nonlocal Cauchy problem:

$$\begin{cases} u'(t) = Au(t) + \displaystyle\int_0^t B(t-s)u(s)ds + f(t, u(t)), & t \geq 0, \\ u(0) = g(u), \end{cases} \tag{8.6}$$

where $g : BC(\mathbb{R}_+, X) \to X$ satisfying the following conditions:

(Cg1) There exists a continuous and nondecreasing function $L_g : [0, +\infty) \to [0, +\infty)$ such that for each $r \geq 0$ and all $u, v \in BC(\mathbb{R}_+, X)$ with $\|u\|_\infty \leq r, \|v\|_\infty \leq r$

$$\|g(u) - g(v)\| \leq L_g(r)\|u - v\|_\infty.$$

(Cg2) $\displaystyle\sup_{r>0} \left[\frac{\delta r}{M} - \delta r L_g(r) - r L_f(r) \right] > \delta \|g(0)\| + \sup_{t \geq 0} \|f(t, 0)\|.$

Theorem 8.2. *Let conditions (HR), (Cf1)–(Cf2) and (Cg1)–(Cg2) hold. Then Eq. (8.6) has a mild solution $u \in PSABP_{\omega,k}(\mathbb{R}_+, X)$.*

Proof. The proof can be conducted similarly to Theorem 8.1. We define the operator F by

$$\begin{aligned} (Fu)(t) &= \mathcal{R}(t)g(u) + \int_0^t \mathcal{R}(t-s)f(s, u(s))ds \\ &:= F_1(t) + F_2(t), \quad t \geq 0, \end{aligned} \tag{8.7}$$

where $F_1(t) = \mathcal{R}(t)g(u)$ and $F_2(t) = \displaystyle\int_0^t \mathcal{R}(t-s)f(s, u(s))ds.$

For each $u \in PSABP_{\omega,k}(\mathbb{R}_+, X)$, it can be easily checked from conditions (HR) and (Cg2) that $F_1(t) \in PSABP_{\omega,k}(\mathbb{R}_+, X)$ (see also the proof of Theorem 7.3 in Chapter 7). Meanwhile the proof of Theorem 8.1 shows that $F_2(t)$ also belongs to $PSABP_{\omega,k}(\mathbb{R}_+, X)$. Thus, the operator $F : PSABP_{\omega,k}(\mathbb{R}_+, X) \to PSABP_{\omega,k}(\mathbb{R}_+, X)$ is well defined by (8.7).

It follows from the condition (Cg2) that there exists a constant $r > 0$ such that

$$\frac{\delta r}{M} - \delta r L_g(r) - r L_f(r) > \delta \|g(0)\| + \sup_{t \geq 0} \|f(t, 0)\|. \tag{8.8}$$

Let $\mathbb{D}$ be defined in Theorem 8.1 and for each $u \in \mathbb{D}$ and all $t \geq 0$, we have

$$\|(Fu)(t)\|$$

$$\leq M[\|g(0)\| + \|g(u) - g(0)\|]$$

$$+ M \int_0^t e^{-\delta(t-s)}[\|f(s,0)\| + \|f(s,u(s)) - f(s,0)\|]ds$$

$$\leq M[\|g(0)\| + L_g(r)r] + \frac{M}{\delta}\Big[\sup_{s \in \mathbb{R}_+} \|f(s,0)\| + rL_f(r)\Big],$$

which from (8.8) implies that $\|Fu\|_\infty \leq r$ and thus $F(\mathbb{D}) \subseteq \mathbb{D}$.

Next we show that F is a contraction on $\mathbb{D}$. It is known from (8.8) that

$$\frac{\delta r}{M} - \delta r L_g(r) - r L_f(r) > 0.$$

Thus we have

$$M L_g(r) + \frac{M}{\delta} L_f(r) < 1. \tag{8.9}$$

For each $u, v \in \mathbb{D}$ and all $t \geq 0$, we have

$$\|(Fu)(t) - (Fv)(t)\|$$

$$\leq \|\mathcal{R}(t)[g(u) - g(v)]\| + \int_0^t \|\mathcal{R}(t-s)[f(s,u(s)) - f(s,v(s))]\|ds$$

$$\leq M L_g(r)\|u - v\|_\infty + M \int_0^t e^{-\delta(t-s)} L_f(r)\|u(s) - v(s)\|ds$$

$$\leq \Big[M L_g(r) + \frac{M}{\delta} L_f(r)\Big] \|u - v\|_\infty.$$

Hence

$$\|Fu - Fv\|_\infty \leq \Big[M L_g(r) + \frac{M}{\delta} L_f(r)\Big] \|u - v\|_\infty.$$

It follows from (8.9) that F is a contraction on $\mathbb{D}$. Thus, F admits a unique fixed point in $\mathbb{D}$, which is naturally a pseudo S-asymptotically (ω, k)-Bloch periodic mild solution to Eq. (8.6). $\qquad\square$

We also have the following corollary.

Corollary 8.2. *Assume that conditions (HR), (Cf1)–(Cf2) and (Cg1) hold. If $L_f(\cdot) \equiv L_1, L_g(\cdot) \equiv L_2$, then Eq. (8.6) admits a unique mild solution $u \in PSABP_{\omega,k}(\mathbb{R}_+, X)$ provided that*

$$0 < L_1 + \delta L_2 < \frac{\delta}{M}.$$

Proof. Noting that

$$\delta\|g(0)\| + \sup_{s\in\mathbb{R}_+} \|f(s,0)\| < \infty,$$

since $0 < L_1 + \delta L_2 < \frac{\delta}{M}$, there exists a constant $r_0 > 0$ such that for all $r \geq r_0$, we have

$$\frac{\delta r}{M} - rL_1 - \delta rL_2 > \delta\|g(0)\| + \sup_{s\in\mathbb{R}_+} \|f(s,0)\|.$$

It follows from the proof of Theorem 8.2 that there exists a unique mild solution $u \in PSABP_{\omega,k}(\mathbb{R}_+, X)$ to Eq. (8.6). $\qquad\square$

Remark 8.1. We can refer to Andrade *et al.* (2012), Brindle and N'Guérékata (2019c), Ding *et al.* (2008), Hernández and Dos Santos (2006), Zhao *et al.* (2013b) and references therein for some concrete examples satisfying the condition (HR) and growth conditions on functions f and g.

Remark 8.2. We can similarly give some existence results for Eqs. (8.1) and (8.6) without the Lipschitz growth condition on the second variable of f. See for instance, Theorem 7.4 in Chapter 7 by using the space defined in Eq. (1.4) in Chapter 1, or Andrade *et al.* (2012, Theorem 3.3) and Brindle and N'Guérékata (2019c, Theorem 4.8).

Chapter 9

Bloch-Type Periodic Solutions to Semilinear Integral Equations

This chapter is mainly concerned with the existence of generalized Bloch-type periodic mild solutions to the following semilinear integral equation:

$$u(t) = \int_{-\infty}^{t} a(t-s)[Au(s) + f(s, u(s))]ds, \quad t \in \mathbb{R}, \tag{9.1}$$

where $a \in L^1(\mathbb{R}_+)$, $A : D(A) \subseteq X \to X$ is the generator of an integral resolvent family $\{S(t)\}_{t \geq 0}$ defined on a Banach space X, and $f : \mathbb{R} \times X \to X$ is a bounded continuous function satisfying some suitable conditions.

Equations of type (9.1) usually arise in the theory of viscoelastic material behavior. Some typical examples can be found in viscoelastic fluids and heat flow in fading memory materials. Under such circumstances, the operator A typically denotes the Laplacian in $X = L^2(\Omega)$ or the elasticity operator, the Stokes operator or biharmonic Δ^2, etc., equipped with some suitable boundary conditions. The pioneering works on existence (and uniqueness) of (compact) almost automorphic mild solutions to Eq. (9.1) can be referred to Cuevas and Lizama (2009), Henríquez and Lizama (2009), and Lizama and Ponce (2011c). Some other results such as existence (and uniqueness) of (weighted) pseudo almost automorphic mild solutions to Eq. (9.1) can be seen in Chang et al. (2014), Zhang et al. (2013), Zhao et al. (2011) and references therein. In this chapter, we continue to investigate some existence results on generalized Bloch-type periodic mild solutions to Eq. (9.1) as an application.

9.1 Integral Resolvent Family and Linear Integral Equation

We first recall some fundamental results on integral resolvent family which are established in Cuevas and Lizama (2009) and Lizama and Poblete (2007).

Definition 9.1. Let A be a closed linear operator with domain $D(A) \subseteq X$. We say that A is the generator of an integral resolvent if there exist $\varpi \geq 0$ and a strongly continuous function $S : \mathbb{R}_+ \to \mathcal{B}(X)$ such that $\left\{ \frac{1}{\hat{a}(\lambda)} : \text{Re}\lambda > \varpi \right\} \subseteq \rho(A)$ and

$$\left(\frac{1}{\hat{a}(\lambda)} I - A \right)^{-1} x = \int_0^\infty e^{-\lambda t} S(t) x \, dt, \quad \text{Re}\lambda > \varpi, \quad x \in X.$$

In this case, $\{S(t)\}_{t \geq 0}$ is said to be the integral resolvent family generated by the operator A.

Lemma 9.1. *Let $\{S(t)\}_{t \geq 0}$ be the integral resolvent family on X with the generator A. Then the following properties hold:*

(a) *$S(t)D(A) \subseteq D(A)$ and $AS(t)x = S(t)Ax$ for all $x \in D(A)$ and $t \geq 0$.*
(b) *Let $x \in D(A)$ and $t \geq 0$. Then*

$$S(t)x = a(t)x + \int_0^t a(t - s)AS(s)x \, ds.$$

(c) *Let $x \in X$ and $t \geq 0$. Then $\int_0^t a(t - s)S(s)x \, ds \in D(A)$ and*

$$S(t)x = a(t)x + A \int_0^t a(t - s)S(s)x \, ds.$$

In particular, $S(0) = a(0)I$.

Remark 9.1 (see Cuevas and Lizama, 2009). According to the uniqueness of the Laplace transform, the above defined integral resolvent family $S(t)$ with $a(t) \equiv 1$ is the same as a C_0-semigroup whereas $S(t)$ with $a(t) = t$ is reduced to the concept of sine family. If $a(t) = t^{\alpha-1}/\Gamma(\alpha)$ with $\alpha \in (1, 2)$, then the integral resolvent family $S(t)$ corresponds to the α-resolvent family $S_\alpha(t)$ in Araya and Lizama (2008) or the (α, α)-resolvent family $R_\alpha(t)$ in Lizama *et al.* (2016). It is also noted that the integral resolvent family $S(t)$ is a special case of (a, k)-regularized family (Lizama, 2000) and just corresponds to a (a, a)-regularized family, see Definition 1.8 and Lemma 1.17 in Chapter 1.

The linear equation related to Eq. (9.1) is such as

$$u(t) = \int_{-\infty}^t a(t - s)[Au(s) + f(s)] ds, \quad t \in \mathbb{R}. \tag{9.2}$$

We recall the following assumption, see for instance Henríquez and Lizama (2009):

(INT) There exists $\phi \in L^1(\mathbb{R}_+)$ such that $\|S(t)\| \leq \phi(t)$, $t \in \mathbb{R}_+$.

The following results are similar to (compact) almost automorphic solutions to Eq. (9.2) established in Cuevas and Lizama (2009), Henríquez and Lizama (2009), and Lizama and Ponce (2011c).

Theorem 9.1. *Let* $a \in L^1(\mathbb{R}_+)$. *Assume further that the integral resolvent family* $\{S(t)\}_{t\geq 0}$ *generated by* A *satisfies the condition (INT). If* $f \in \mathcal{N}([D(A)])$, *then Eq. (9.2) admits a unique solution* $u \in \mathcal{N}([D(A)])$, *which is given by to the equation*

$$u(t) = \int_{-\infty}^{t} S(t-s)f(s)ds, \quad t \in \mathbb{R}.$$

Proof. Let $u(t)$ be defined by

$$u(t) = \int_{-\infty}^{t} S(t-s)f(s)ds, \quad t \in \mathbb{R}.$$

For a given $f \in \mathcal{N}([D(A)])$, we can conclude that the above given $u \in \mathcal{N}([D(A)])$ from Theorems 2.1, 2.2, 2.7, 2.10, 2.13 of Chapter 2, respectively, due to the condition (INT). It follows by Lemma 9.1(b) and the Fubini theorem that

$$\int_{-\infty}^{t} a(t-s)Au(s)ds$$

$$= \int_{-\infty}^{t} a(t-s)A \int_{-\infty}^{s} S(s-\tau)f(\tau)d\tau ds$$

$$= \int_{-\infty}^{t} \int_{-\infty}^{s} a(t-s)AS(s-\tau)f(\tau)d\tau ds$$

$$= \int_{-\infty}^{t} \int_{\tau}^{t} a(t-s)AS(s-\tau)f(\tau)dsd\tau$$

$$= \int_{-\infty}^{t} \int_{0}^{t-\tau} a(t-\tau-p)AS(p)dpf(\tau)d\tau$$

$$= \int_{-\infty}^{t} (S(t-\tau)f(\tau) - a(t-\tau)f(\tau))d\tau$$

$$= u(t) - \int_{-\infty}^{t} a(t-\tau)f(\tau))d\tau,$$

which implies u is a solution to Eq. (9.2). $\qquad\square$

Particularly, let $X = \mathbb{R}$, we have the following results for the scalar equation. Similar results can be found in Cuevas and Lizama (2009) and Henríquez and Lizama (2009).

Corollary 9.1. *Let $f \in \mathcal{N}(\mathbb{R})$, $a \in L^1(\mathbb{R}_+)$, and let $\varrho > 0$ be a real number. If the solution $S_\varrho(t)$ to the following one-dimensional equation*

$$S_\varrho(t) = a(t) - \varrho \int_0^t a(t-s)S_\varrho(s)ds, \tag{9.3}$$

satisfies $|S_\varrho(t)| \leq \phi_\varrho(t)$, where $\phi_\varrho \in L^1(\mathbb{R}_+)$, then the equation

$$u(t) = \int_{-\infty}^t a(t-s)[-\varrho u(s) + f(s)]ds, \quad t \in \mathbb{R}, \tag{9.4}$$

has a solution $u \in \mathcal{N}(\mathbb{R})$ given by

$$u(t) = \int_{-\infty}^t S_\varrho(t-s)f(s)ds, \quad t \in \mathbb{R}. \tag{9.5}$$

Corollary 9.2. *Let $f \in \mathcal{N}(\mathbb{R})$ and $\varrho > 0$ be a real number. If $a \in L^1(\mathbb{R}_+)$ is positive, nonincreasing and log-convex, then the following properties are fulfilled:*

(a) *There exists $S_\varrho \in L^1(\mathbb{R}_+) \cap C(\mathbb{R}_+)$ satisfying Eq. (9.3).*
(b) *Equation (9.4) admits a solution $u \in \mathcal{N}(\mathbb{R})$ given by Eq. (9.5).*

Proof. Assertion (a) follows from (Prüss, 1993, Lemma 4.1, p. 98), and assertion (b) is a consequence of (a) and Corollary 9.1.   □

Remark 9.2. It is noted that if $a \in L^1(\mathbb{R}_+)$ is positive, nonincreasing and log-convex, then a is completely positive (Clément and Prato, 1998).

9.2   Semilinear Integral Equation

In this section, we mainly consider some existence results of generalized Bloch-type periodic mild solutions to Eq. (9.1).

Definition 9.2. Let the operator A be the generator of an integral resolvent family $\{S(t)\}_{t \geq 0}$. A continuous function $u : \mathbb{R} \to X$ is said to be a mild solution to Eq. (9.1) if it satisfies the following integral equation:

$$u(t) = \int_{-\infty}^t S(t-s)f(s, u(s))ds, \quad t \in \mathbb{R}.$$

Theorem 9.2. *Assume that the operator A generates an integral resolvent family $\{S(t)\}_{t\geq 0}$ satisfying the condition (INT). Suppose further that $f = g + h \in BC(\mathbb{R} \times X, X)$ satisfies the Lipschitz-type condition*

$$\|f(t,x) - f(t,y)\| \leq L\|x - y\|, \tag{9.6}$$

for all $t \in \mathbb{R}$ and each $x, y \in X$, where $0 < L < \|\phi\|_{L^1(\mathbb{R}_+)}^{-1}$ and g satisfies the condition (A1) in Section 2.3 of Chapter 2, $h \in \mathscr{E}(\mathbb{R} \times X, X)$. Then Eq. (9.1) has a unique mild solution $u \in PBP_{\omega,k}(\mathbb{R}, X)$.

Proof. Taking into account Theorem 9.1 and Corollary 2.2(1) of Chapter 2, we can complete the proof analogously to that of Theorem 4.5 of Chapter 4. We omit details here. □

As a direct consequence of Corollary 9.2 and Theorem 9.2, we have the following result for the scalar equation.

Corollary 9.3. *Let $\varrho > 0$ be a real number. Assume that $a \in L^1(\mathbb{R}_+)$ is a positive, nonincreasing and log-convex function. If $f = g + h \in BC(\mathbb{R} \times \mathbb{R}, \mathbb{R})$ satisfies the Lipschitz-type condition (9.6) for all $t, x, y \in \mathbb{R}$ with g satisfying the condition (A1) in Section 2.3 of Chapter 2, $h \in \mathscr{E}(\mathbb{R} \times \mathbb{R}, \mathbb{R})$, then there exists $S_\varrho \in L^1(\mathbb{R}_+) \cap C(\mathbb{R}_+)$ verifying Eq. (9.3). Moreover, if $0 < L < \|S_\rho\|_{L^1(\mathbb{R}_+)}^{-1}$, then the semilinear equation*

$$u(t) = \int_{-\infty}^{t} a(t-s)[-\varrho u(s) + f(s, u(s))]ds, \ t \in \mathbb{R},$$

admits a unique solution $u \in PBP_{\omega,k}(\mathbb{R}, \mathbb{R})$.

Different Lipschitz-type conditions (see, e.g., Cuevas and Lizama, 2009) are considered in the following results.

Theorem 9.3. *Let the operator A generate a uniformly bounded integral resolvent family $\{S(t)\}_{t\geq 0}$ satisfying the condition (INT). Let $f = g + h \in BC(\mathbb{R} \times X, X)$ verify the following condition:*

$$\|f(t,x) - f(t,y)\| \leq L_f(t)\|x - y\|, \tag{9.7}$$

for all $t \in \mathbb{R}$ and each $x, y \in X$, where $L_f \in L^1(\mathbb{R}, \mathbb{R}_+)$, g satisfies the condition (A1) in Section 2.3 of Chapter 2 and $h \in \mathscr{E}(\mathbb{R} \times X, X)$. Then Eq. (9.1) has a unique mild solution $u \in PBP_{\omega,k}(\mathbb{R}, X)$.

Proof. Since the integral resolvent family $\{S(t)\}_{t\geq 0}$ is uniformly bounded, there exists $N > 0$ such that $\|S(t)\| \leq N$ for all $t \geq 0$. Taking into account Theorem 9.1 and Corollary 2.3 of Chapter 2, we can complete the proof similarly to that of Theorem 4.6. We omit details here. □

An immediate consequence of Corollary 9.2 and Theorem 9.3 is the following result.

Corollary 9.4. *Let $\varrho > 0$ be a real number. Assume that $a \in L^1(\mathbb{R}_+)$ is a positive, nonincreasing and* log-convex *function. If $f = g + h \in BC(\mathbb{R} \times \mathbb{R}, \mathbb{R})$ satisfies the condition* (9.7) *for all $t, x, y \in \mathbb{R}$ with g satisfying the condition (A1) in Section 2.3 of Chapter 2, $h \in \mathscr{E}(\mathbb{R} \times \mathbb{R}, \mathbb{R})$, then there exists $S_\varrho \in L^1(\mathbb{R}_+) \cap C(\mathbb{R}_+)$ verifying Eq. (9.3). Moreover, if $\{S_\varrho(t)\}_{t \geq 0}$ is uniformly bounded, then the semilinear equation*

$$u(t) = \int_{-\infty}^{t} a(t - s)[-\varrho u(s) + f(s, u(s))]ds, \quad t \in \mathbb{R},$$

admits a unique solution $u \in PBP_{\omega,k}(\mathbb{R}, \mathbb{R})$.

Theorem 9.4. *Assume that the integral resolvent family $\{S(t)\}_{t \geq 0}$ generated by A is uniformly exponentially stable. Let $f = g + h \in BC(\mathbb{R} \times X, X)$ verify the condition* (9.7) *for all $t \in \mathbb{R}$ and each $x, y \in X$, where $L_f \in L^p(\mathbb{R}, \mathbb{R}_+)$ $(1 \leq p < +\infty)$, g satisfies the condition (A1) in Section 2.3 of Chapter 2 and $h \in \mathscr{E}(\mathbb{R} \times X, X)$. Then Eq. (9.1) has a unique mild solution $u \in PBP_{\omega,k}(\mathbb{R}, X)$.*

Proof. Since $\{S(t)\}_{t \geq 0}$ is uniformly exponentially stable, there exist constants $M, \delta > 0$ such that $\|S(t)\| \leq Me^{-\delta t}$ for all $t \geq 0$. Thus, the condition (INT) is satisfied. Taking into account Theorem 9.1 and Corollary 2.3 of Chapter 2, we can complete the proof similarly to that of Theorem 3.6 of Chapter 3. We omit details here.   □

Remark 9.3. For concrete examples of such operators which are uniformly exponentially stable, we can refer to Prüss (2009, Section 5).

Theorem 9.5. *Let the operator A generate an integral resolvent family $\{S(t)\}_{t \geq 0}$ satisfying $\|S(t)\| \leq \phi(t)$ for all $t \geq 0$, where $\phi : \mathbb{R}_+ \to \mathbb{R}_+$ is a decreasing function such that $\phi_0 := \sum_{m=0}^{\infty} \phi(m) < \infty$. Assume further that $f = g + h \in BC(\mathbb{R} \times X, X)$ verify the condition* (9.7) *for all $t \in \mathbb{R}$ and each $x, y \in X$, where $L_f \in L^1(\mathbb{R}, \mathbb{R}_+)$, g satisfies the condition (A1) in Section 2.3 of Chapter 2 and $h \in \mathscr{E}(\mathbb{R} \times X, X)$. Then Eq. (9.1) has a unique mild solution $u \in PBP_{\omega,k}(\mathbb{R}, X)$ whenever*

$$\bar{L}\phi_0 < 1 \text{ with } \bar{L} := \sup_{t \in \mathbb{R}} \int_{t}^{t+1} L_f(s)ds.$$

Proof. The proof can be conducted similarly to Cuevas and Lizama (2009, Theorem 4.6). Since ϕ is decreasing and $\phi_0 := \sum_{m=0}^{\infty} \phi(m) < \infty$, we conclude that $\phi \in L^1(\mathbb{R}_+)$ and thus ϕ is uniformly integrable. Define the operator $F : PBP_{\omega,k}(\mathbb{R}, X) \to PBP_{\omega,k}(\mathbb{R}, X)$ by

$$(F\Phi)(t) := \int_{-\infty}^{t} S(t-s)f(s, \Phi(s))ds, \quad t \in \mathbb{R}.$$

For each $\Phi \in PBP_{\omega,k}(\mathbb{R}, X)$, Corollary 2.3 of Chapter 2 implies that $f(s, \Phi(s)) \in PBP_{\omega,k}(\mathbb{R}, X)$. Thus, $F\Phi \in PBP_{\omega,k}(\mathbb{R}, X)$ and F is well defined according to Theorem 9.1. Let $u, v \in PBP_{\omega,k}(\mathbb{R}, X)$, we have

$$\|(Fu)(t) - (Fv)(t)\|$$

$$= \left\| \int_{-\infty}^{t} S(t-s)[f(s, u(s)) - f(s, v(s))]ds \right\|$$

$$\leq \int_{-\infty}^{t} L_f(s)\|S(t-s)\|\|u(s) - v(s)\|ds$$

$$\leq \left(\sum_{m=0}^{\infty} \int_{t-(m+1)}^{t-m} L_f(s)\phi(t-s)ds \right) \|u-v\|_{\infty}$$

$$\leq \left(\sum_{m=0}^{\infty} \phi(m) \int_{t-(m+1)}^{t-m} L_f(s)ds \right) \|u-v\|_{\infty}$$

$$\leq \bar{L}\phi_0 \|u-v\|_{\infty},$$

i.e., $\|Fu - Fv\|_{\infty} \leq \bar{L}\phi_0\|u-v\|_{\infty}$. We can complete the proof by the Banach fixed point theorem whenever $\bar{L}\phi_0 < 1$. $\qquad \square$

Remark 9.4. Let $\rho \in \mathcal{U}_{inv}$ and take into account Theorem 9.1, Corollaries 2.2(2) and 2.4 of Chapter 2, we can obtain Eq. (9.1) admits a unique mild solution $u \in WPBP_{\omega,k}(\mathbb{R}, X, \rho)$ such as Theorems 9.2–9.5.

Theorem 9.6. *Let the operator A generate an integral resolvent family $\{S(t)\}_{t\geq 0}$ satisfying the condition (INT). Suppose further that $f \in BC(\mathbb{R} \times X, X)$ satisfies conditions (9.6) and (T1) in Theorem 2.5 of Chapter 2 for all $t \in \mathbb{R}$ and each $x, y \in X$ with $0 < L < \|\phi\|_{L^1(\mathbb{R}_+)}^{-1}$. Then Eq. (9.1) has a unique mild solution $u \in SABP_{\omega,k}(\mathbb{R}, X)$.*

Proof. The proof is mainly based upon Theorems 9.1 and 2.5 of Chapter 2, and can be conducted analogously to that of Theorem 4.5 of Chapter 4. We omit details here. $\qquad \square$

The following corollary is a direct consequence of Corollary 9.2 and Theorem 9.6.

Corollary 9.5. *Let $\varrho > 0$ be a real number. Assume that $a \in L^1(\mathbb{R}_+)$ is a positive, nonincreasing and log-convex function. If $f \in BC(\mathbb{R} \times \mathbb{R}, \mathbb{R})$ satisfies the condition (T1) in Theorem 2.5 of Chapter 2 and the Lipschitz-type condition (9.6) for all $t, x, y \in \mathbb{R}$, then there exists $S_\varrho \in L^1(\mathbb{R}_+) \cap C(\mathbb{R}_+)$ verifying Eq. (9.3). Moreover, if $0 < L < \|S_\rho\|_{L^1(\mathbb{R}_+)}^{-1}$, then the semilinear equation*

$$u(t) = \int_{-\infty}^t a(t-s)[-\varrho u(s) + f(s, u(s))]ds, \quad t \in \mathbb{R},$$

admits a unique solution $u \in SABP_{\omega,k}(\mathbb{R}, \mathbb{R})$.

Remark 9.5. Taking into account Theorems 9.1, 2.8, and 2.11 of Chapter 2, we can obtain that Eq. (9.1) has a unique mild solution $u \in PSABP_{\omega,k}(\mathbb{R}, X)$ (or $WPSABP_{\omega,k}(\mathbb{R}, X, \rho)$) such as Theorem 9.6, respectively.

Next we investigate the existence of pseudo (ω, k)-Bloch periodic mild solutions to (9.1) when f is not necessarily Lipschitz with its second variable but has the following growth condition (F2). Similar results on existence of compact almost automorphic or (weighted) pseudo almost automorphic mild solutions to (9.1) can be found in Chang *et al.* (2014), Henríquez and Lizama (2009), Zhang *et al.* (2013), and Zhao *et al.* (2011). We shall use the space defined in Eq. (1.5) of Chapter 1.

Theorem 9.7. *Assume that A generates an integral resolvent family $\{S(t)\}_{t \geq 0}$ satisfying the condition (INT) and $f = g + h \in BC(\mathbb{R} \times X, X)$ with g satisfying the condition (A1) in Section 2.3 of Chapter 2, $h \in \mathscr{E}(\mathbb{R} \times X, X)$. Suppose further that:*

(F1) *$f \in BC(\mathbb{R} \times X, X)$ and $f(t, x)$ is uniformly continuous in any bounded subset $\mathbb{K} \subseteq X$ uniformly for $t \in \mathbb{R}$.*

(F2) *There exists a continuous nondecreasing function $W_f : \mathbb{R}_+ \to \mathbb{R}_+$ such that*

$$\|f(t, x)\| \leq W_f(\|x\|) \quad \text{for all} \quad t \in \mathbb{R} \quad \text{and} \quad x \in X.$$

(F3) *For each $r \geq 0$, the function $t \to \int_{-\infty}^t \phi(t-s)W_f(rh(s))ds$ belongs to $BC(\mathbb{R})$. We set*

$$\beta(r) = \left\| \int_{-\infty}^t \phi(t-s)W_f(rh(s))ds \right\|_h.$$

(F4) *For each $\varepsilon > 0$ there is $\delta > 0$ such that for every $u, v \in C_{\mathbb{h}}(X)$, $\|u - v\|_{\mathbb{h}} \leq \delta$ implies that*

$$\int_{-\infty}^{t} \phi(t - s)\|f(s, u(s)) - f(s, v(s))\| ds \leq \varepsilon$$

for all $t \in \mathbb{R}$.

(F5) $\lim_{\xi \to \infty} \frac{\xi}{\beta(\xi)} > 1.$

(F6) *For all $a, b \in \mathbb{R}, a < b$ and $r > 0$, the set $\{f(s, x) : a \leq s \leq b,\ x \in C_{\mathbb{h}}(X), \|x\|_{\mathbb{h}} \leq r\}$ is relatively compact in X.*

Then Eq. (9.1) has at least one mild solution $u \in PBP_{\omega,k}(\mathbb{R}, X)$.

Proof. We define the linear operator $\Lambda : C_{\mathbb{h}}(X) \to C_{\mathbb{h}}(X)$ by

$$(\Lambda u)(t) := \int_{-\infty}^{t} S(t - s)f(s, u(s))\, ds, \quad t \in \mathbb{R}.$$

We will show that Λ has a fixed point in $PBP_{\omega,k}(\mathbb{R}, X)$. For the sake of convenience, we divide the proof into several steps.

(I) For each $u \in C_{\mathbb{h}}(X)$, it follows by conditions (INT) and (F2) that

$$\|(\Lambda u)(t)\| \leq \int_{-\infty}^{t} \phi(t - s)W_f(\|u(s)\|)ds$$

$$\leq \int_{-\infty}^{t} \phi(t - s)W_f(\|u\|_{\mathbb{h}}\mathbb{h}(s))ds.$$

Thus, the condition (F3) implies that $\Lambda : C_{\mathbb{h}}(X) \to C_{\mathbb{h}}(X)$.

(II) The operator Λ is continuous. In fact, for any $\varepsilon > 0$, we take $\delta > 0$ involved in the condition (F4). If $u, v \in C_{\mathbb{h}}(X)$ and $\|u - v\|_{\mathbb{h}} \leq \delta$, then

$$\|(\Lambda u)(t) - (\Lambda v)(t)\| \leq \int_{-\infty}^{t} \phi(t - s)\|f(s, u(s)) - f(s, v(s))\| ds \leq \varepsilon,$$

which shows the assertion.

(III) We will show that Λ is completely continuous. We set $B_r(Z)$ for the closed ball with center at 0 and radius r in the space Z. Let $V = \Lambda(B_r(C_{\mathbb{h}}(X)))$ and $v = \Lambda(u)$ for $u \in B_r(C_{\mathbb{h}}(X))$. Firstly, we will prove that $V(t)$ is a relatively compact subset of X for each $t \in \mathbb{R}$. It follows from the condition (F3) that the function $s \to \phi(s)W_f(r\mathbb{h}(t - s))$ is integrable on $\mathbb{R}_+$. Hence, for $\varepsilon > 0$, we can choose $a \geq 0$ such that $\int_{a}^{\infty} \phi(s)W_f(r\mathbb{h}(t - s))ds \leq \varepsilon$. Since

$$v(t) = \int_{0}^{a} S(s)f(t - s, u(t - s))\, ds + \int_{a}^{\infty} S(s)f(t - s, u(t - s))\, ds$$

and

$$\left\| \int_a^\infty S(s)f\left(t-s, u(t-s)\right) ds \right\| \le \int_a^\infty \phi(s)W_f(r\mathbb{h}(t-s))ds \le \varepsilon,$$

we get $v(t) \in a\overline{co(K)} + B_\varepsilon(X)$, where $co(K)$ denotes the convex hull of K and $K = \{S(s)f(\xi, u) : 0 \le s \le a,\ t - a \le \xi \le t,\ \|u\|_\mathbb{h} \le r\}$. Using the strong continuity of $S(\cdot)$ and the condition (F6), we infer that K is a relatively compact set, and $V(t) \subseteq a\overline{co(K)} + B_\varepsilon(X)$, which establishes our assertion.

Secondly, we show that the set V is equicontinuous. In fact, we can decompose

$$v(t+s) - v(t) = \int_0^s S(\sigma)f\left(t+s-\sigma, u(t+s-\sigma)\right) d\sigma$$
$$+ \int_0^a [S(\sigma+s) - S(\sigma)]f\left(t-\sigma, u(t-\sigma)\right) d\sigma$$
$$+ \int_a^\infty [S(\sigma+s) - S(\sigma)]f\left(t-\sigma, u(t-\sigma)\right) d\sigma.$$

For each $\varepsilon > 0$, we can choose $a > 0$ and $\delta_1 > 0$ such that

$$\left\| \int_0^s S(\sigma)f\left(t+s-\sigma, u(t+s-\sigma)\right) d\sigma \right.$$
$$\left. + \int_a^\infty [S(\sigma+s) - S(\sigma)]f\left(t-\sigma, u(t-\sigma)\right) d\sigma \right\|$$
$$\le \int_0^s \phi(\sigma)W_f(r\mathbb{h}(t+s-\sigma))d\sigma$$
$$+ \int_a^\infty [\phi(\sigma+s) + \phi(\sigma)]W_f(r\mathbb{h}(t-\sigma))d\sigma$$
$$\le \frac{\varepsilon}{2}$$

for $s \le \delta_1$. Moreover, since $\{f\left(t-\sigma, u(t-\sigma)\right) : 0 \le \sigma \le a,\ u \in B_r\left(C_\mathbb{h}(X)\right)\}$ is a relatively compact set and $S(\cdot)$ is strongly continuous, we can choose $\delta_2 > 0$ such that for $s \le \delta_2$

$$\|[S(\sigma+s) - S(\sigma)]f\left(t-\sigma, u(t-\sigma)\right)\| \le \frac{\varepsilon}{2a}.$$

Combining these estimates, we get $\|v(t+s)-v(t)\| \le \varepsilon$ for s sufficiently small and independent of $u \in B_r\left(C_\mathbb{h}(X)\right)$.

Finally, applying the condition (F3), it is easy to see that

$$\frac{\|v(t)\|}{\mathrm{h}(t)} \leq \frac{1}{\mathrm{h}(t)} \int_{-\infty}^{t} \phi(t-s) W_f(r\mathrm{h}(s)) ds \to 0, \quad |t| \to \infty,$$

and this convergence is independent of $u \in B_r\left(C_{\mathrm{h}}(X)\right)$. Hence, by Lemma 1.20 of Chapter 1, V is a relatively compact set in $C_{\mathrm{h}}(X)$.

(IV) Let $u^{\lambda}(\cdot)$ be a solution to $u^{\lambda} = \lambda\Lambda(u^{\lambda})$ for some $0 < \lambda < 1$. We can estimate

$$\|u^{\lambda}(t)\| = \lambda \left\| \int_{-\infty}^{t} S(t-s) f\left(s, u^{\lambda}(s)\right) ds \right\|$$

$$\leq \int_{-\infty}^{t} \phi(t-s) W_f(\|u^{\lambda}\|_{\mathrm{h}}\mathrm{h}(s)) ds$$

$$\leq \beta(\|u^{\lambda}\|_{\mathrm{h}})\mathrm{h}(t).$$

Hence, we have

$$\frac{\|u^{\lambda}\|_{\mathrm{h}}}{\beta(\|u^{\lambda}\|_{\mathrm{h}})} \leq 1$$

and combining with the condition (F5), we conclude that the set $\{u^{\lambda} : u^{\lambda} = \lambda\Lambda(u^{\lambda}),\ \lambda \in (0,1)\}$ is bounded.

(V) By Corollary 2.8(1) of Chapter 2 and Theorem 9.1, we have $\Lambda\left(PBP_{\omega,k}(\mathbb{R}, X)\right) \subseteq PBP_{\omega,k}(\mathbb{R}, X)$. Noting that $PBP_{\omega,k}(\mathbb{R}, X)$ is a closed subspace of $C_{\mathrm{h}}(X)$, hence we can consider $\Lambda : \overline{PBP_{\omega,k}(\mathbb{R}, X)} \to \overline{PBP_{\omega,k}(\mathbb{R}, X)}$. By steps (I)-(III), we deduce that this map is completely continuous. Applying Lemma 1.22 of Chapter 1 we infer that Λ has a fixed point $u \in \overline{PBP_{\omega,k}(\mathbb{R}, X)}$. Let $\{u_n\}$ be a sequence in $PBP_{\omega,k}(\mathbb{R}, X)$ which converges to u by the norm of $C_{\mathrm{h}}(X)$, we have

$$\|\Lambda u^n - u\|_{\infty} = \|\Lambda u^n - \Lambda u\|_{\infty}$$

$$\leq \sup_{t \in \mathbb{R}} \frac{1}{\mathrm{h}(t)} \int_{0}^{t} \phi(t-s) \|f\left(s, u^n(s)\right) - f(s, u(s))\| ds.$$

It follows from the condition (F4) that $\Lambda u^n \to u$, as $n \to \infty$ uniformly in $\mathbb{R}$. Since $\Lambda u^n \in PBP_{\omega,k}(\mathbb{R}, X)$, we have $u \in PBP_{\omega,k}(\mathbb{R}, X)$, i.e., (9.1) admits one mild solution $u \in PBP_{\omega,k}(\mathbb{R}, X)$. $\square$

Remark 9.6. Let $\rho \in \mathcal{U}_{inv}$ and take into account Theorem 9.1, Corollary 2.8(2) of Chapter 2, we can obtain that Eq. (9.1) admits at least one mild solution $u \in WPBP_{\omega,k}(\mathbb{R}, X, \rho)$ such as Theorems 9.7.

Theorem 9.8. *Let the operator A generate an integral resolvent family $\{S(t)\}_{t\geq 0}$ satisfying the condition (INT). Suppose that $f \in BC(\mathbb{R} \times X, X)$ satisfies the condition (T1) in Theorem 2.5 of Chapter 2. Assume further that conditions (F1)–(F6) hold. Then Eq. (9.1) admits at least one mild solution $u \in SABP_{\omega,k}(\mathbb{R}, X)$.*

Proof. Taking into account Theorems 2.6 and 9.1, we can complete the proof similarly to that of Theorem 9.7. We omit details here. $\square$

Remark 9.7. Taking into account Theorems 2.9, 2.12, of Chapter 2 and Theorem 9.1, we can obtain Eq. (9.1) has at least one mild solution $u \in PSABP_{\omega,k}(\mathbb{R}, X, \rho)$ (or $WPSABP_{\omega,k}(\mathbb{R}, X, \rho)$) such as Theorem 9.8, respectively.

Appendix A

Compactness of Fractional Resolvent Operator Families

In this chapter, we list basic results on norm continuity (continuity in $\mathcal{B}(X)$) and characterizations of compactness for some fractional resolvent operator families which appear in Chapters 4 and 5.

A.1   Norm Continuity and Compactness of the Resolvent Operator $S_{\alpha,\beta}(t)$

In this section, we give some results on the norm continuity and characterizations of compactness for resolvent operator $S_{\alpha,\beta}(t)$ for given $\alpha, \beta > 0$ and all $t > 0$. These properties with applications can be founded in Chang and Ponce (2021), Chang *et al.* (2021), Fan (2014), Lizama *et al.* (2016), and Ponce (2016, 2020b).

We recall that the (α, β)-resolvent operator family $\{S_{\alpha,\beta}(t)\}_{t \geq 0}$ is given in Definition 4.1 of Chapter 4 and has some basic properties such as Proposition 4.1 and Theorem 4.1 of Chapter 4. For $\mu > 0$, we define

$$g_\mu(t) = \begin{cases} \dfrac{t^{\mu-1}}{\Gamma(\mu)}, & t > 0, \\ 0, & t \leq 0, \end{cases}$$

where $\Gamma(\cdot)$ is the Gamma function. We also define $g_0 \equiv \delta_0$, the Dirac delta. For $\alpha > 0$, let $n := \lceil \alpha \rceil$ denote the smallest integer greater than or equal to α. Recall that the Caputo and Riemann–Liouville fractional derivative of order α for u are defined, respectively, by

$$\partial_c^\alpha u(t) := \int_0^t g_{n-\alpha}(t-s) u^{(n)}(s) ds,$$

and

$$\partial_r^\alpha u(t) := \frac{d^n}{dt^n} \int_0^t g_{n-\alpha}(t-s) u(s) ds.$$

Moreover, we have for $1 < \alpha < 2$ that

$$\begin{cases} \widehat{\partial_c^\alpha u}(\lambda) = \lambda^\alpha \hat{u}(\lambda) - \lambda^{\alpha-1}u(0) - \lambda^{\alpha-2}u'(0), \\ \widehat{\partial_r^\alpha u}(\lambda) = \lambda^\alpha \hat{u}(\lambda) - \lambda(g_{2-\alpha} * u)(0) - (g_{2-\alpha} * u)'(0). \end{cases} \tag{A.1}$$

Let $BUC(\mathbb{R}_+, X)$ be the space of all bounded, uniformly continuous functions $f : \mathbb{R}_+ \to X$. We first list the following lemmas.

Lemma A.1 (Weis, 1988, Corollary 2.3). *Let (Ω, μ) be a measure space and $t \in \Omega \to T_t \in \mathcal{B}(X, Y)$ be a strongly integrable function, i.e.,*

$$Tx = \int_\Omega T_t x d\mu(t) \tag{A.2}$$

exists for all $x \in X$ as a Bochner integral and $\int_\Omega \|T_t\| d\mu(t) < \infty$. If μ-almost all T_t in (A.2) are compact, then T is compact.

Lemma A.2 (Haase, 2008, Proposition 2.1). *Let X, Y be Banach spaces, let $\mathbb{S} : [0, \infty) \to \mathcal{B}(X, Y)$ be strongly continuous, and let $a \in L_{loc}^1[0, \infty)$ be a scalar function, both a and $\mathbb{S}$ of finite exponential type. Then for every $\omega > \omega_0(\mathbb{S}), \omega_0(a)$ one has*

$$\lim_{N \to \infty} \frac{1}{2\pi i} \int_{\omega-iN}^{\omega+iN} e^{\lambda t} \widehat{(a * \mathbb{S})}(\lambda) d\lambda = a * \mathbb{S},$$

in $\mathcal{B}(X, Y)$, uniformly in t from compact subsets of $[0, \infty)$, where $\omega_0(\mathbb{S}) := \inf\{\nu \in \mathbb{R} : \|\mathbb{S}\| \leq Me^{\nu t}(t \geq 0)\}$.

Lemma A.3 (Arendt et al., 2001, Proposition 1.3.5). *Let $f \in L_{loc}^1(\mathbb{R}_+, X)$ and $\mathcal{T} : \mathbb{R}_+ \to \mathcal{B}(X, Y)$ be strongly continuous.*

(a) *(Young's inequality) Let $1 \leq p, q, r \leq \infty$ satisfy $1/p + 1/q = 1 + 1/r$. If $\int_0^\infty \|\mathcal{T}(t)\|^p dt < \infty$ and $f \in L^q(\mathbb{R}_+, X)$, then $\mathcal{T} * f \in L^r(\mathbb{R}_+, Y)$ and*

$$\|\mathcal{T} * f\|_r \leq \|f\|_q \left(\int_0^\infty \|\mathcal{T}(t)\|^p dt \right)^{1/p}.$$

(b) *Let $1 < p, p' < \infty$ satisfy $1/p + 1/p' = 1$. If $\int_0^\infty \|\mathcal{T}(t)\|^p dt < \infty$ and $f \in L^{p'}(\mathbb{R}_+, X)$, then $\mathcal{T} * f \in C_0(\mathbb{R}_+, Y)$.*

(c) *If $\int_0^\infty \|\mathcal{T}(t)\| dt < \infty$ and $f \in BUC(\mathbb{R}_+, X)$, or $\mathcal{T}$ is bounded and $f \in L^1(\mathbb{R}_+, X)$, then $\mathcal{T} * f \in BUC(\mathbb{R}_+, Y)$.*

(d) *If $\int_0^\infty \|T(t)\| dt < \infty$ and $f \in C_0(\mathbb{R}_+, X)$, or $\lim_{t\to\infty} \|T(t)\| = 0$ and $f \in L^1(\mathbb{R}_+, X)$, then $\mathcal{T} * f \in C_0(\mathbb{R}_+, Y)$.*

Theorem A.1 (Ponce, 2016, Proposition 11). *Let $\alpha > 0$ and $1 < \beta \le 2$. Suppose that $\{S_{\alpha,\beta}(t)\}_{t \ge 0}$ is the (α, β)-resolvent family of type (M, δ) generated by the operator A. Then the function $t \mapsto S_{\alpha,\beta}(t)$ is continuous in $\mathcal{B}(X)$ for all $t > 0$.*

Proof. Firstly, let $1 < \beta < 2$. It is noticed that, for all $\text{Re}\lambda > 0$

$$\widehat{\left(S_{\alpha,\beta}\right)}(\lambda) = \lambda^{\alpha-\beta}(\lambda^\alpha - A)^{-1}$$

$$= \frac{1}{\lambda^{\beta-1}}\lambda^{\alpha-1}(\lambda^\alpha - A)^{-1}$$

$$= \widehat{\left(g_{\beta-1} * S_{\alpha,1}\right)}(\lambda).$$

By the uniqueness of the Laplace transform, it is concluded that $S_{\alpha,\beta}(t) = (g_{\beta-1} * S_{\alpha,1})(t)$, for all $t > 0$. Now take $0 < t_0 < t_1$. Then

$$S_{\alpha,\beta}(t_1) - S_{\alpha,\beta}(t_0)$$

$$= (g_{\beta-1} * S_{\alpha,1})(t_1) - (g_{\beta-1} * S_{\alpha,1})(t_0)$$

$$= \int_{t_0}^{t_1} g_{\beta-1}(t_1 - r)S_{\alpha,1}(r)dr$$

$$+ \int_0^{t_0} [g_{\beta-1}(t_1 - r) - g_{\beta-1}(t_0 - r)]S_{\alpha,1}(r)dr$$

$$=: I_1 + I_2.$$

Since $\beta > 1$, we obtain $g_\beta(0) = 0$ and

$$\|I_1\| \le \int_{t_0}^{t_1} g_{\beta-1}(t_1 - r)\|S_{\alpha,1}(r)\|dr$$

$$\le Me^{\delta t_1}g_\beta(t_1 - t_0) \to 0,$$

and thus $\|I_1\| \to 0$ as $t_1 \to t_0$.

On the other side, we have

$$\|I_2\| \le \int_0^{t_0} |g_{\beta-1}(t_1 - r) - g_{\beta-1}(t_0 - r)|\|S_{\alpha,1}(r)\|dr$$

$$\le Me^{\delta t_1}\int_0^{t_0} |g_{\beta-1}(t_1 - r) - g_{\beta-1}(t_0 - r)|dr$$

$$= Me^{\delta t_1}\int_0^{t_0} |g_{\beta-1}(t_1 - t_0 + r) - g_{\beta-1}(r)|dr.$$

Since $1 < \beta < 2$, the function $r \mapsto g_{\beta-1}(r)$ is decreasing in $\mathbb{R}_+$ and thus $g_{\beta-1}(r) - g_{\beta-1}(t_1 - t_0 + r) > 0$, for all $r > 0$, obtaining

$$\|I_2\| \leq Me^{\delta t_1}[g_\beta(t_0) - g_\beta(t_1) + g_\beta(t_1 - t_0)] \to 0, \text{ as } t_1 \to t_0.$$

Hence we conclude that $S_{\alpha,\beta}(t)$ is norm continuous for $1 < \beta < 2$.

Finally, if $\beta = 2$, then the uniqueness of the Laplace transform implies

$$S_{\alpha,2}(t)x = (g_1 * S_{\alpha,1})(t)x = \int_0^t S_{\alpha,1}(r)x \, dr,$$

for all $x \in X$. For $0 < t_0 < t_1$ we have

$$\|S_{\alpha,2}(t_1)x - S_{\alpha,2}(t_0)x\|$$

$$\leq \int_{t_0}^{t_1} \|S_{\alpha,1}(r)x\| dr \leq Me^{\delta t_1}\|x\|(t_1 - t_0) \to 0,$$

as $t_1 \to t_0$ for all $x \in X$. Thus $\|S_{\alpha,2}(t_1) - S_{\alpha,2}(t_0)\| \to 0$ as $t_1 \to t_0$. $\square$

Theorem A.2 (Ponce, 2016, Lemma 12). *Suppose that the operator A generates an (α, β)-resolvent family $\{S_{\alpha,\beta}(t)\}_{t\geq 0}$ of type (M, δ). If $\gamma > 0$, then A also generates an $(\alpha, \beta + \gamma)$-resolvent family of type $\left(\frac{M}{\delta^\gamma}, \delta\right)$.*

Proof. It follows from hypothesis that for all $t > 0$,

$$\|(g_\gamma * S_{\alpha,\beta})(t)\| \leq M \int_0^t g_\gamma(t - s)e^{\delta s} ds$$

$$\leq Me^{\delta t} \int_0^t g_\gamma(s)e^{-\delta s} ds$$

$$\leq Me^{\delta t} \int_0^\infty g_\gamma(s)e^{-\delta s} ds$$

$$= \frac{Me^{\delta t}}{\delta^\gamma}.$$

Thus, $(g_\gamma * S_{\alpha,\beta})(t)$ is Laplace transformable and, for all $\text{Re}\lambda > \delta$, we have

$$\widehat{\left(g_\gamma * S_{\alpha,\beta}\right)}(\lambda) = \frac{1}{\lambda^\gamma}\lambda^{\alpha-\beta}(\lambda^\alpha - A)^{-1}$$

$$= \lambda^{\alpha-(\beta+\gamma)}(\lambda^\alpha - A)^{-1}$$

$$= \widehat{\left(S_{\alpha,\beta+\gamma}\right)}(\lambda).$$

It is concluded that the operator A generates an $(\alpha, \beta + \gamma)$-resolvent family of type $\left(\frac{M}{\delta^\gamma}, \delta\right)$. $\square$

In what follows, we will assume that $\{S_{\alpha,\beta}(t)\}_{t\geq 0}$ is strongly continuous for all $\alpha, \beta > 0$.

Theorem A.3 (Ponce, 2016, Theorem 14). *Let $\alpha > 0$, $1 < \beta \leq 2$ and the operator A generate an (α, β)-resolvent family $\{S_{\alpha,\beta}(t)\}_{t\geq 0}$ of type (M, δ). Then the following assertions are equivalent:*

(i) $S_{\alpha,\beta}(t)$ *is a compact operator for all $t > 0$.*
(ii) $(\mu - A)^{-1}$ *is a compact operator for all $\mu > \delta^{1/\alpha}$.*

Proof. (i) $\Rightarrow$ (ii) Suppose that the resolvent family $\{S_{\alpha,\beta}(t)\}_{t>0}$ is compact. Let $\lambda > \delta$ be fixed. Then we have

$$\lambda^{\alpha-\beta}(\lambda^{\alpha} - A)^{-1} = \int_0^{\infty} e^{-\lambda t} S_{\alpha,\beta}(t) dt,$$

where the integral in the right-hand side exists in the Bochner sense. Since $\{S_{\alpha,\beta}(t)\}_{t>0}$ is continuous in $\mathcal{B}(X)$ (see Theorem A.1) we conclude that $(\lambda^{\alpha} - A)^{-1}$ is a compact operator by by Lemma A.1.

(ii) $\Rightarrow$ (i) Conversely, let $t > 0$ be fixed and $1 < \beta < 2$. Since $\beta > 1$, it follows that $g_{\beta-1} \in L^1_{\mathrm{loc}}[0, \infty)$ and thus, by Lemma A.2 we obtain

$$\lim_{N\to\infty} \frac{1}{2\pi i} \int_{\delta-iN}^{\delta+iN} e^{\lambda t} \widehat{(g_{\beta-1} * S_{\alpha,1})}(\lambda) d\lambda$$

$$= (g_{\beta-1} * S_{\alpha,1})(t) = S_{\alpha,\beta}(t),$$

in $\mathcal{B}(X)$. Hence,

$$\frac{1}{2\pi i} \int_{\Xi} e^{\lambda t} \lambda^{\alpha-\beta} (\lambda^{\alpha} - A)^{-1} d\lambda = S_{\alpha,\beta}(t), \ t > 0,$$

where Ξ is the path consisting of the vertical line $\{\delta + is : s \in \mathbb{R}\}$. By hypothesis and Lemma A.1, we conclude that $S_{\alpha,\beta}(t)$ is compact for all $\alpha > 0$ and $1 < \beta < 2$. Now, in case $\beta = 2$ we observe that in $\mathcal{B}(X)$ we have

$$\lim_{N\to\infty} \frac{1}{2\pi i} \int_{\delta-iN}^{\delta+iN} e^{\lambda t} \widehat{(g_1 * S_{\alpha,1})}(\lambda) d\lambda$$

$$= (g_1 * S_{\alpha,1})(t) = S_{\alpha,2}(t),$$

by Lemma A.2, and we conclude that $S_{\alpha,2}(t)$ is compact for all $t > 0$, analogously to case $1 < \beta < 2$. $\square$

We have the following corollary by Theorem A.3.

Corollary A.1 (Ponce, 2016, Corollary 15). *Let $1 < \alpha \leq 2$ and $\{S_{\alpha,\alpha}(t)\}_{t\geq 0}$ be an (α, α)-resolvent family of type (M, δ) generated by the operator A. Then the following assertions are equivalent:*

(i) $S_{\alpha,\alpha}(t)$ is a compact operator for all $t > 0$.

(ii) $(\mu - A)^{-1}$ is a compact operator for all $\mu > \delta^{1/\alpha}$.

To get a compactness criterion of $\{S_{\alpha,1}(t)\}_{t \geq 0}$ and $\{S_{\alpha,\alpha-1}(t)\}_{t \geq 0}$ (i.e., in case $\beta = 1$ and $\beta = \alpha - 1$) we need an additional hypothesis: the norm continuity of $t \mapsto S_{\alpha,1}(t)$ and $t \mapsto S_{\alpha,\alpha-1}(t)$, respectively.

Theorem A.4 (Ponce, 2016, Proposition 16). *Let $1 < \alpha < 2$, and $\{S_{\alpha,1}(t)\}_{t \geq 0}$ be the $(\alpha, 1)$-resolvent family of type (M, δ) generated by the operator A. Suppose that $S_{\alpha,1}(t)$ is continuous in the uniform operator topology for all $t > 0$. Then the following assertions are equivalent:*

(i) $S_{\alpha,1}(t)$ *is a compact operator for all $t > 0$.*

(ii) $(\mu - A)^{-1}$ *is a compact operator for all $\mu > \delta^{1/\alpha}$.*

Proof. (i) $\Rightarrow$ (ii) Suppose that the resolvent family $\{S_{\alpha,1}(t)\}_{t > 0}$ is compact. Let $\lambda > \delta$ be fixed. Then we have

$$\lambda^{\alpha-1}(\lambda^\alpha - A)^{-1} = \int_0^\infty e^{-\lambda t} S_{\alpha,1}(t) dt,$$

where the integral in the right-hand side exists in the Bochner sense, because $\{S_{\alpha,1}(t)\}_{t > 0}$ is continuous in the uniform operator topology, by hypothesis. Then, by Lemma A.1 we conclude that $(\lambda^\alpha - A)^{-1}$ is a compact operator.

(ii) $\Rightarrow$ (i) Let $t > 0$ be fixed. Since $1 < \alpha < 2$, it follows that $g_{2-\alpha} \in L^1_{\text{loc}}[0, \infty)$ and therefore, by Lemma A.2 we obtain

$$\lim_{N \to \infty} \frac{1}{2\pi i} \int_{\delta - iN}^{\delta + iN} e^{\lambda t} \widehat{(g_{2-\alpha} * S_{\alpha,\alpha-1})}(\lambda) d\lambda$$

$$= (g_{2-\alpha} * S_{\alpha,\alpha-1})(t) = S_{\alpha,1}(t),$$

in $\mathcal{B}(X)$. Therefore,

$$\frac{1}{2\pi i} \int_\Xi e^{\lambda t} \lambda^{\alpha-1}(\lambda^\alpha - A)^{-1} d\lambda = S_{\alpha,1}(t), \quad t > 0,$$

where Ξ is the path consisting of the vertical line $\{\delta + is : s \in \mathbb{R}\}$. By hypothesis and Lemma A.1, we conclude that $S_{\alpha,1}(t)$ is compact. $\square$

Theorem A.5 (Ponce, 2016, Proposition 17). *Let $3/2 < \alpha < 2$. Assume that $\{S_{\alpha,\alpha-1}(t)\}_{t \geq 0}$ is the $(\alpha, \alpha-1)$-resolvent family of type (M, δ) generated by A, and $S_{\alpha,\alpha-1}(t)$ is continuous in the uniform operator topology for all $t > 0$. Then the following assertions are equivalent:*

(i) $S_{\alpha,\alpha-1}(t)$ *is a compact operator for all* $t > 0$.
(ii) $(\mu - A)^{-1}$ *is a compact operator for all* $\mu > \delta^{1/\alpha}$.

Proof. (i) $\Rightarrow$ (ii) It follows as in the proof of Theorem A.4.

(ii) $\Rightarrow$ (i) Let $t > 0$ be fixed. Since $\alpha > 3/2$, it follows that $g_{\alpha-\frac{3}{2}} \in L^1_{\text{loc}}[0, \infty)$ and therefore, by Lemma A.2 we obtain

$$\lim_{N \to \infty} \frac{1}{2\pi i} \int_{\delta-iN}^{\delta+iN} e^{\lambda t} \widehat{\left(g_{\alpha-\frac{3}{2}} * S_{\alpha,\frac{1}{2}}\right)}(\lambda) d\lambda$$
$$= \left(g_{\alpha-\frac{3}{2}} * S_{\alpha,\frac{1}{2}}\right)(t) = S_{\alpha,\alpha-1}(t),$$

in $\mathcal{B}(X)$. Therefore,

$$\frac{1}{2\pi i} \int_{\Xi} e^{\lambda t} \lambda^{\alpha-1} (\lambda^\alpha - A)^{-1} d\lambda = S_{\alpha,\alpha-1}(t),$$

where Ξ is the path consisting of the vertical line $\{\delta + is : s \in \mathbb{R}\}$. By hypothesis and Lemma A.1, we conclude that $S_{\alpha,\alpha-1}(t)$ is compact. $\square$

Theorem A.6 (Ponce, 2016, Proposition 18). *Let* $1/2 < \alpha < 1$. *Assume that* $\{S_{\alpha,\alpha}(t)\}_{t \geq 0}$ *is the* (α, α)-*resolvent family of type* (M, δ) *generated by the operator* A, *and* $S_{\alpha,\alpha}(t)$ *is continuous in the uniform operator topology for all* $t > 0$. *Then, the following assertions are equivalent:*

(i) $S_{\alpha,\alpha}(t)$ *is a compact operator for all* $t > 0$.
(ii) $(\mu - A)^{-1}$ *is a compact operator for all* $\mu > \delta^{1/\alpha}$.

Proof. It can be conducted similarly to Theorem A.4, since for $1/2 < \alpha < 1$ we have

$$\lim_{N \to \infty} \frac{1}{2\pi i} \int_{\delta-iN}^{\delta+iN} e^{\lambda t} \widehat{\left(g_{\alpha-\frac{1}{2}} * S_{\alpha,\frac{1}{2}}\right)}(\lambda) d\lambda$$
$$= \left(g_{\alpha-\frac{1}{2}} * S_{\alpha,\frac{1}{2}}\right)(t) = S_{\alpha,\alpha}(t),$$

in $\mathcal{B}(X)$ and $t > 0$ by Lemma A.2. $\square$

Remark A.1. (Ponce, 2016, Remark 19). Let $\varepsilon_0 > 0$ be fixed and $\varepsilon_0 < \alpha < 1$. It follows by Lemma A.2 that

$$\lim_{N \to \infty} \frac{1}{2\pi i} \int_{\delta-iN}^{\delta+iN} e^{\lambda t} \widehat{\left(g_{\alpha-\varepsilon_0} * S_{\alpha,\varepsilon_0}\right)}(\lambda) d\lambda$$
$$= \left(g_{\alpha-\varepsilon_0} * S_{\alpha,\varepsilon_0}\right)(t) = S_{\alpha,\alpha}(t),$$

in $\mathcal{B}(X)$. Therefore, as is Theorem A.6, if $\alpha > \varepsilon_0$ with $\varepsilon_0 > 0$, the operator A generates the (α, α)-resolvent family $\{S_{\alpha,\alpha}(t)\}_{t\geq 0}$ of type (M, δ) and $S_{\alpha,\alpha}(t)$ is norm continuous for all $t > 0$, then $S_{\alpha,\alpha}(t)$ is a compact operator for all $t > 0$ if and only if $(\lambda^\alpha - A)^{-1}$ is a compact operator for all $\lambda > \delta^{1/\alpha}$. Meanwhile, if $\varepsilon_0 < \alpha < 2$ with $\varepsilon_0 > 1$ being fixed and $\{S_{\alpha,\alpha-1}(t)\}_{t\geq 0}$ is the $(\alpha, \alpha - 1)$-resolvent family of type (M, δ) generated by the operator A, which is norm continuous for all $t > 0$, then $S_{\alpha,\alpha-1}(t)$ is a compact operator for all $t > 0$ if and only if $(\lambda^\alpha - A)^{-1}$ is a compact operator for all $\lambda > \delta^{1/\alpha}$.

A.2   Norm Continuity and Compactness of the Resolvent Operator $S_{\alpha,\beta}^E(t)$

In this section, we give some results on the norm continuity and characterizations of compactness for resolvent operator $S_{\alpha,\beta}^E(t)$ for given $\alpha > 0, 1 < \beta \leq 2$ and all $t > 0$. This kind of resolvent operator $S_{\alpha,\beta}^E(t)$ is usually applied to deal with fractional evolution equations of Sobolev (or degenerate) type, see for instance Chang *et al.* (2017), Chang and Ponce (2019), Chang *et al.* (2019b), Chang and Pei (2020), Pei and Chang (2020), and Ponce (2020d).

The *E-modified resolvent set of* A, $\rho_E(A)$, defined by
$$\rho_E(A) := \{\lambda \in \mathbb{C} : (\lambda E - A) : D(A) \cap D(E) \to X \text{ is invertible and } (\lambda E - A)^{-1} \in \mathcal{B}(X, [D(A) \cap D(E)])\}.$$
The operator $(\lambda E - A)^{-1}$ is called the *E-resolvent operator of* A.

Definition A.1. Let $A : D(A) \subseteq X \to X$, $E : D(E) \subseteq X \to X$ be closed linear operators defined on a Banach space X satisfying $D(A) \cap D(E) \neq \{0\}$. Let $\alpha, \beta > 0$. We say that the pair (A, E) is the generator of an (α, β)-resolvent family, if there exist $\widetilde{\omega} \geq 0$ and a strongly continuous function $S_{\alpha,\beta}^E : [0, \infty) \to \mathcal{B}([D(E)], X)$ such that $S_{\alpha,\beta}^E(t)$ is exponentially bounded, $\{\lambda^\alpha : \text{Re}\lambda > \widetilde{\omega}\} \subset \rho_E(A)$, and for all $x \in D(E)$,

$$\lambda^{\alpha-\beta}\left(\lambda^\alpha E - A\right)^{-1} Ex = \int_0^\infty e^{-\lambda t} S_{\alpha,\beta}^E(t) x \, dt, \quad \text{Re}\lambda > \widetilde{\omega}.$$

In this case, $\left\{S_{\alpha,\beta}^E(t)\right\}_{t\geq 0}$ is called the (α, β)-*resolvent family* generated by the pair (A, E).

Remark A.2. It is easy to show from Lemmas 1.16 and 1.17 of Chapter 1 (see also Lizama (2000, Proposition 3.1 and Lemma 2.2)) that if the pair (A, E) generates an (α, β)-resolvent family $\left\{S_{\alpha,\beta}^E(t)\right\}_{t\geq 0}$, then it satisfies the following properties:

(i) $ES^E_{\alpha,\beta}(0)x = g_\beta(0)Ex$, for all $x \in D(E)$;

(ii) $ES^E_{\alpha,\beta}(t)x = g_\beta(t)Ex + \int_0^t g_\alpha(t-s)AS^E_{\alpha,\beta}(s)xds$, for all $x \in D(E)$ and $t \geq 0$;

It is noticed that the Definition 4.1 of Chapter 4 corresponds to some well-known concepts in the literature. In fact, the case $S^E_{1,1}(t)$ corresponds to degenerate semigroups (see Favini and Yagi, 1999; Melnikova and Filinkov, 2001, Chapter 1, Section 1.5), if $\alpha = 1, \beta = k+1$, then $S^E_{1,k+1}(t)$ is a degenerated k-integrated semigroup (see Melnikova and Filinkov, 2001, Chapter 1, Section 1.5; Arendt and Favini, 1993), and $S^E_{2,1}(t)$ is a cosine degenerate family (see Melnikova and Filinkov, 2001, Chapter 1, Section 1.7). If $E = I$, then $S^I_{1,1}(t), S^I_{1,k+1}(t)$ and $S^I_{2,1}(t)$ correspond to a C_0-semigroup, a k-integrated semigroup and a cosine family, respectively. Finally, if $\beta = 1$, then $S^I_{\alpha,1}(t)$ is the α-resolvent family (also called the α-times resolvent family) for fractional differential equations (see Fan, 2014).

Let $1 < \alpha < 2$ and A, E be closed linear operators defined on X. Consider the Caputo fractional Cauchy problem

$$\begin{cases} \partial_c^\alpha(Eu)(t) = Au(t) + Ef(t), & t \geq 0, \\ Eu(0) = Ex, \\ (Eu)'(0) = Ey. \end{cases} \tag{A.3}$$

Taking Laplace transform in (A.3) and by Definition A.1, we can see that the mild solution to Eq. (A.3) can be given as

$$u(t) = S^E_{\alpha,1}(t)x + S^E_{\alpha,2}(t)y + \int_0^t S^E_{\alpha,\alpha}(t-s)f(s)ds, \quad t \geq 0.$$

Furthermore, if the pair (A, E) generates an $(\alpha, 1)$-resolvent operator $S^E_{\alpha,1}(t)$, then by the uniqueness of the Laplace transform, the mild solution of Eq. (A.3) can be expressed by

$$u(t) = S^E_{\alpha,1}(t)x + (g_1 * S^E_{\alpha,1})(t)y$$
$$+ \int_0^t (g_{\alpha-1} * S^E_{\alpha,1})(t-s)f(s)ds, \quad t \geq 0. \tag{A.4}$$

Similarly, for the Riemann–Liouville fractional Cauchy problem

$$\begin{cases} \partial_r^\alpha(Eu)(t) = Au(t) + Ef(t), & t \geq 0, \\ (E(g_{2-\alpha} * u))(0) = Ex, \\ (E(g_{2-\alpha} * u))'(0) = Ey, \end{cases} \tag{A.5}$$

if the pair (A, E) generates an $(\alpha, \alpha - 1)$-resolvent operator $S^E_{\alpha,\alpha-1}(t)$, then by the uniqueness of the Laplace transform, the mild solution of Eq. (A.5) can be expressed by

$$
u(t) = S^E_{\alpha,\alpha-1}(t)x + S^E_{\alpha,\alpha}(t)y + \int_0^t S^E_{\alpha,\alpha}(t-s)f(s)ds
$$

$$
= S^E_{\alpha,\alpha-1}(t)x + \left(g_1 * S^E_{\alpha,\alpha-1}\right)(t)y
$$

$$
+ \int_0^t \left(g_1 * S^E_{\alpha,\alpha-1}\right)(t-s)f(s)ds, \quad t \geq 0. \tag{A.6}
$$

In what follows, we list some results on the norm continuity and compactness of $S^E_{\alpha,\beta}(t)$ for given $\alpha > 0, 1 < \beta \leq 2$, which are similar to ones in Chang *et al.* (2017).

Theorem A.7 (Chang *et al.*, 2017, Proposition 3.1). *Let* $\alpha > 0$ *and* $1 < \beta \leq 2$. *Suppose that* $\left\{S^E_{\alpha,\beta}(t)\right\}_{t \geq 0}$ *is the* (α, β)-*resolvent family of type* (M, δ) *generated by the pair* (A, E). *Then the function* $t \mapsto S^E_{\alpha,\beta}(t)$ *is continuous in* $\mathcal{B}(X)$ *for all* $t > 0$.

Proof. First, let $1 < \beta < 2$. By the uniqueness of the Laplace transform, it is obvious to see that $S^E_{\alpha,\beta}(t) = \left(g_{\beta-1} * S^E_{\alpha,1}\right)(t)$, for all $t > 0$. Now, we take $0 < t_0 < t_1$. Then

$$
S^E_{\alpha,\beta}(t_1) - S^E_{\alpha,\beta}(t_0)
$$

$$
= \left(g_{\beta-1} * S^E_{\alpha,1}\right)(t_1) - \left(g_{\beta-1} * S^E_{\alpha,1}\right)(t_0)
$$

$$
= \int_{t_0}^{t_1} g_{\beta-1}(t_1 - r)S^E_{\alpha,1}(r)dr
$$

$$
+ \int_0^{t_0} [g_{\beta-1}(t_1 - r) - g_{\beta-1}(t_0 - r)]S^E_{\alpha,1}(r)dr
$$

$$
=: I_1 + I_2.
$$

Since $\beta > 1$, $g_\beta(0) = 0$ and we obtain

$$
\|I_1\| \leq \int_{t_0}^{t_1} g_{\beta-1}(t_1 - r)\|S^E_{\alpha,1}(r)\|dr
$$

$$
\leq Me^{\delta t_1}g_\beta(t_1 - t_0) \to 0, \quad \text{as } t_1 \to t_0.
$$

On the other hand,

$$\|I_2\| \le \int_0^{t_0} |g_{\beta-1}(t_1 - r) - g_{\beta-1}(t_0 - r)| \|S_{\alpha,1}^E(r)\| dr$$

$$\le M e^{\delta t_1} \int_0^{t_0} |g_{\beta-1}(t_1 - r) - g_{\beta-1}(t_0 - r)| dr$$

$$= M e^{\delta t_1} \int_0^{t_0} |g_{\beta-1}(t_1 - t_0 + r) - g_{\beta-1}(r)| dr.$$

Since $1 < \beta < 2$, the function $r \mapsto g_{\beta-1}(r)$ is decreasing in $[0, \infty)$ and therefore $g_{\beta-1}(r) - g_{\beta-1}(t_1 - t_0 + r) > 0$, for all $r > 0$, obtaining

$$\|I_2\| \le M e^{\delta t_1} [g_\beta(t_0) - g_\beta(t_1) + g_\beta(t_1 - t_0)] \to 0, \text{ as } t_1 \to t_0.$$

Therefore, $S_{\alpha,\beta}^E(t)$ is continuous for $1 < \beta < 2$.

Finally, if $\beta = 2$, then the uniqueness of the Laplace transform implies

$$S_{\alpha,2}^E(t)x = (g_1 * S_{\alpha,1}^E)(t)x = \int_0^t S_{\alpha,1}^E(r)x dr,$$

for all $x \in X$. For $0 < t_0 < t_1$ we have

$$\|S_{\alpha,2}^E(t_1)x - S_{\alpha,2}^E(t_0)x\|$$

$$\le \int_{t_0}^{t_1} \|S_{\alpha,1}^E(r)x\| dr \le M e^{\delta t_1} \|x\| (t_1 - t_0) \to 0,$$

as $t_1 \to t_0$, for all $x \in X$. Thus, $\|S_{\alpha,2}^E(t_1) - S_{\alpha,2}^E(t_0)\| \to 0$ as $t_1 \to t_0$. □

Theorem A.8 (Chang et al., 2017, Lemma 3.1). *Suppose that the pair (A, E) generates an (α, β)-resolvent family $\{S_{\alpha,\beta}^E(t)\}_{t \ge 0}$ of type (M, δ). If $\gamma > 0$, then the pair (A, E) also generates an $(\alpha, \beta + \gamma)$-resolvent family of type $(\frac{M}{\delta^\gamma}, \delta)$.*

Proof. Analogous to the proof of Theorem A.2. By hypothesis we get for all $t \ge 0$,

$$\|(g_\gamma * S_{\alpha,\beta}^E)(t)\| \le M \int_0^t g_\gamma(t - s) e^{\delta s} ds \le M e^{\delta t} \int_0^t g_\gamma(s) e^{-\delta s} ds$$

$$\le M e^{\delta t} \int_0^\infty g_\gamma(s) e^{-\delta s} ds = \frac{M e^{\delta t}}{\delta^\gamma}.$$

Therefore, $(g_\gamma * S_{\alpha,\beta}^E)(t)$ is Laplace transformable and for all $\lambda > \delta$, we have

$$\widehat{(g_\gamma * S_{\alpha,\beta}^E)}(\lambda)$$

$$= \frac{1}{\lambda^\gamma} \lambda^{\alpha-\beta} (\lambda^\alpha E - A)^{-1} E = \lambda^{\alpha-(\beta+\gamma)} (\lambda^\alpha E - A)^{-1} E$$

$$= \widehat{(S_{\alpha,\beta+\gamma}^E)}(\lambda).$$

It is concluded that the pair (A, E) generates an $(\alpha, \beta + \gamma)$-resolvent family of type $\left(\frac{M}{\delta^\gamma}, \delta\right)$. $\qquad\square$

Theorem A.9 (Chang *et al.*, 2017, Theorem 3.1). *Let $\alpha > 0$, $1 < \beta \le 2$. Assume that $\{S_{\alpha,\beta}^E(t)\}_{t>0}$ is an (α, β)-resolvent family of type (M, δ) generated by the pair (A, E). Then the following assertions are equivalent:*

(i) $S_{\alpha,\beta}^E(t)$ *is a compact operator for all $t > 0$.*
(ii) $(\mu E - A)^{-1} E$ *is a compact operator for all $\mu > \delta^{1/\alpha}$.*

Proof. (i) $\Rightarrow$ (ii) Assume that the resolvent family $\{S_{\alpha,\beta}^E(t)\}_{t>0}$ is compact. Let $\lambda > \delta$ be fixed. Then we have

$$\lambda^{\alpha-\beta}(\lambda^\alpha E - A)^{-1} E = \int_0^\infty e^{-\lambda t} S_{\alpha,\beta}^E(t)\, dt,$$

where the integral in the right-hand side exists in the Bochner sense, since $\{S_{\alpha,\beta}^E(t)\}_{t>0}$ is continuous in the uniform operator topology (by Theorem A.7), we conclude that $(\lambda^\alpha E - A)^{-1} E$ is a compact operator by Lemma A.1.

(ii) $\Rightarrow$ (i) Let $t > 0$ be fixed. Assume that $1 < \beta < 2$. Since $\beta > 1$, it follows that $g_{\beta-1} \in L^1_{\text{loc}}[0, \infty)$ and therefore, by Lemma A.2 we obtain

$$\lim_{N \to \infty} \frac{1}{2\pi i} \int_{\delta-iN}^{\delta+iN} e^{\lambda t} \widehat{\left(g_{\beta-1} * S_{\alpha,1}^E\right)}(\lambda)\, d\lambda$$

$$= \left(g_{\beta-1} * S_{\alpha,1}^E\right)(t) = S_{\alpha,\beta}^E(t),$$

in $\mathcal{B}(X)$. Hence,

$$\frac{1}{2\pi i} \int_{\Xi} e^{\lambda t} \lambda^{\alpha-\beta} \left(\lambda^\alpha E - A\right)^{-1} E\, d\lambda = S_{\alpha,\beta}^E(t), \ t > 0,$$

where Ξ is the path consisting of the vertical line $\{\delta + is : s \in \mathbb{R}\}$. It is concluded by hypothesis and Lemma A.1 that $S_{\alpha,\beta}^E(t)$ is compact for all $\alpha > 0$ and $1 < \beta < 2$. Now, in case $\beta = 2$ we observe that in $\mathcal{B}(X)$ we have

$$\lim_{N \to \infty} \frac{1}{2\pi i} \int_{\delta-iN}^{\delta+iN} e^{\lambda t} \widehat{\left(g_1 * S_{\alpha,1}^E\right)}(\lambda)\, d\lambda = (g_1 * S_{\alpha,1}^E)(t) = S_{\alpha,2}^E(t),$$

by Lemma A.2, and we conclude that $S_{\alpha,2}^E(t)$ is compact for all $t > 0$, analogously to case $1 < \beta < 2$. $\qquad\square$

Theorem A.9 implies the following corollary.

Corollary A.2 (Chang *et al.*, 2017, Corollary 3.1). *Let $1 < \alpha \le 2$ and $\{S_{\alpha,\alpha}^E(t)\}_{t\ge0}$ be an (α, α)-resolvent family of type (M, δ) generated by the pair (A, E). Then the following assertions are equivalent:*

(i) $S_{\alpha,\alpha}^E(t)$ is a compact operator for all $t > 0$.

(ii) $(\mu E - A)^{-1}E$ is a compact operator for all $\mu > \delta^{1/\alpha}$.

To get a compactness criterion of $\{S_{\alpha,1}^E(t)\}_{t \geq 0}$ and $\{S_{\alpha,\alpha-1}^E(t)\}_{t \geq 0}$ (i.e., in case $\beta = 1$ and $\beta = \alpha - 1$), we need an additional condition: the norm continuity of $t \mapsto S_{\alpha,1}^E(t)$ and $t \mapsto S_{\alpha,\alpha-1}^E(t)$, respectively. Again, the proofs follow similarly to Theorems A.5, A.6 and A.9.

Theorem A.10 (Chang *et al.*, 2017, Proposition 3.2). *Let* $1 < \alpha < 2$. *Assume that* $\{S_{\alpha,1}^E(t)\}_{t \geq 0}$ *is the* $(\alpha, 1)$-*resolvent family of type* (M, δ) *generated by the pair* (A, E). *Suppose further that* $S_{\alpha,1}^E(t)$ *is continuous in the uniform operator topology for all* $t > 0$. *Then the following assertions are equivalent:*

(i) $S_{\alpha,1}^E(t)$ *is a compact operator for all* $t > 0$.

(ii) $(\mu E - A)^{-1}E$ *is a compact operator for all* $\mu > \delta^{1/\alpha}$.

Proof. (i) $\Rightarrow$ (ii) Suppose that the resolvent family $\{S_{\alpha,1}^E(t)\}_{t>0}$ is compact. Let $\lambda > \delta$ be fixed. Then we have

$$\lambda^{\alpha-1}(\lambda^\alpha E - A)^{-1}E = \int_0^\infty e^{-\lambda t}S_{\alpha,1}^E(t)dt,$$

where the integral in the right-hand side exists in the Bochner sense, because $\{S_{\alpha,1}^E(t)\}_{t>0}$ is continuous in the uniform operator topology, by hypothesis. Then, by Lemma A.1 we conclude that $(\lambda^\alpha E - A)^{-1}E$ is a compact operator.

(ii) $\Rightarrow$ (i) Let $t > 0$ be fixed. Since $1 < \alpha < 2$, it follows that $g_{2-\alpha} \in L^1_{\text{loc}}[0, \infty)$ and therefore, by Lemma A.2 we obtain

$$\lim_{N \to \infty} \frac{1}{2\pi i} \int_{\delta-iN}^{\delta+iN} e^{\lambda t}\big(\widehat{g_{2-\alpha} * S_{\alpha,\alpha-1}^E}\big)(\lambda)d\lambda$$

$$= \big(g_{2-\alpha} * S_{\alpha,\alpha-1}^E\big)(t) = S_{\alpha,1}^E(t),$$

in $\mathcal{B}(X)$. Therefore,

$$\frac{1}{2\pi i} \int_\Xi e^{\lambda t}\lambda^{\alpha-1}(\lambda^\alpha E - A)^{-1}Ed\lambda = S_{\alpha,1}^E(t), \quad t > 0,$$

where Ξ is the path consisting of the vertical line $\{\delta + is : s \in \mathbb{R}\}$. By hypothesis and Lemma A.1, we conclude that $S_{\alpha,1}^E(t)$ is compact. $\qquad\square$

The proof of the next result follows the same lines of Theorem A.9. We omit the details.

Theorem A.11 (Chang *et al.*, 2017, Proposition 3.3). *Let $3/2 < \alpha < 2$. Assume that $\{S^E_{\alpha,\alpha-1}(t)\}_{t\geq 0}$ is the $(\alpha, \alpha - 1)$-resolvent family of type (M, δ) generated by the pair (A, E). Suppose further that $S^E_{\alpha,\alpha-1}(t)$ is continuous in the uniform operator topology for all $t > 0$. Then the following assertions are equivalent:*

(i) $S^E_{\alpha,\alpha-1}(t)$ *is a compact operator for all $t > 0$.*
(ii) $(\mu E - A)^{-1}E$ *is a compact operator for all $\mu > \delta^{1/\alpha}$.*

Remark A.3. We can use another differently defined resolvent operator family $\{S^E_{\alpha,\beta}(t)\}_{t\geq 0}$ in Chang *et al.* (2017, Definition 2.4) to deal with the following Cauchy problems

$$\begin{cases} \partial^\alpha_c (Eu)(t) = Au(t) + f(t), & t \geq 0, \\ Eu(0) = x, \\ (Eu)'(0) = y, \end{cases} \tag{A.7}$$

and

$$\begin{cases} \partial^\alpha_r (Eu)(t) = Au(t) + f(t), & t \geq 0, \\ (E(g_{2-\alpha} * u))(0) = x, \\ (E(g_{2-\alpha} * u))'(0) = y. \end{cases} \tag{A.8}$$

Under such circumstances, the mild solutions to problems (A.7) and (A.8) can still be formally expressed by (A.4) and (A.6), respectively. However, the resolvent operator family $\{S^E_{\alpha,\beta}(t)\}_{t\geq 0}$ satisfies the following properties:

(i) $ES^E_{\alpha,\beta}(0)x = g_\beta(0)x$, for all $x \in X$;
(ii) $ES^E_{\alpha,\beta}(t)x = g_\beta(t)x + \int_0^t g_\alpha(t-s)AS^E_{\alpha,\beta}(s)xds$, for all $x \in X$ and $t \geq 0$.

A.3 Applications

As applications of resolvent operator family $\{S_{\alpha,\beta}(t)\}_{t\geq 0}$, in this section we investigate the asymptotic behavior of mild solutions to abstract fractional Cauchy problems for the Caputo and Riemann–Liouville fractional derivatives. We can refer to Ponce (2020b) for details.

Let A be a closed linear operator defined in X, $x, y \in X$, and $1 < \alpha < 2$. First, we consider the initial value problem for the Caputo fractional derivative

$$\begin{cases} \partial^\alpha_c u(t) = Au(t) + f(t), & t \geq 0, \\ u(0) = x, \\ u'(0) = y. \end{cases} \tag{A.9}$$

Applying the Laplace transform to (A.9) we obtain by (A.1) that

$$u(t) = S_{\alpha,1}(t)x + S_{\alpha,2}(t)y + \int_0^t S_{\alpha,\alpha}(t-s)f(s)ds, \quad t \geq 0. \quad (A.10)$$

By a mild solution to the problem (A.9) it is understood that a function $u : [0, \infty) \to X$ satisfying (A.10).

Similarly, for the Riemann–Liouville fractional derivative, if we take Laplace in the problem

$$\begin{cases} \partial_r^\alpha u(t) = Au(t) + f(t), & t \geq 0, \\ (g_{2-\alpha} * u)(0) = x, \\ (g_{2-\alpha} * u)'(0) = y, \end{cases} \quad (A.11)$$

then

$$u(t) = S_{\alpha,\alpha-1}(t)x + S_{\alpha,\alpha}(t)y + \int_0^t S_{\alpha,\alpha}(t-s)f(s)ds, \quad t \geq 0. \quad (A.12)$$

Analogously, a mild solution to the problem (A.11) is a function $u : [0, \infty) \to X$ satisfying (A.12).

Proposition A.1 (Ponce, 2020b, Proposition 4.10). *Let $1 < \alpha < 2$, $\varpi < 0$ and A is ϖ-sectorial of angle $\theta = \frac{(\alpha-1)}{2}\pi$ defined on a Banach space $(X, \| \cdot \|)$. If $f \in L^1(\mathbb{R}_+, X)$ and $y \in X$, then the mild solution u to the problem*

$$\begin{cases} \partial_r^\alpha u(t) = Au(t) + f(t), & t \geq 0, \\ (g_{2-\alpha} * u)(0) = 0, \\ (g_{2-\alpha} * u)'(0) = y, \end{cases} \quad (A.13)$$

belongs to $C_0(\mathbb{R}_+, X)$.

Proof. The mild solution to the problem (A.13) is given by

$$u(t) = S_{\alpha,\alpha}(t)y + \int_0^t S_{\alpha,\alpha}(t-s)f(s)ds, \quad t \geq 0. \quad (A.14)$$

Theorems 4.1 and 4.2 of Chapter 4 show that the family $\{S_{\alpha,\alpha}(t)\}_{t \geq 0}$ is strongly continuous. By Corollary 4.1(a) of Chapter 4 and Lemma A.3(d) (see also Arendt *et al.*, 2001, Proposition 1.3.5) it follows that the convolution $S_{\alpha,\alpha} * f$ belongs to $C_0(\mathbb{R}_+, X)$. Then it follows by Corollary 4.1(a) of Chapter 4 that

$$\|u(t)\| \leq \|S_{\alpha,\alpha}(t)\| \, \|y\| + \|(S_{\alpha,\alpha} * f)(t)\| \to 0$$

as $t \to \infty$. $\qquad \square$

In the next result, we consider a nonzero vector in the first initial condition.

Proposition A.2 (Ponce, 2020b, Proposition 4.11). *Let $1 < \alpha < 2$, $\varpi < 0$ and A is ϖ-sectorial of angle $\theta = \frac{(\alpha-1)}{2}\pi$ defined on a Banach space $(X, \|\cdot\|)$. If $f \in L^1(\mathbb{R}_+, X)$, then the mild solution to the problem (A.9) belongs to $C_0(\mathbb{R}_+, X)$.*

Proof. The mild solution to (A.9) is given by

$$u(t) = S_{\alpha,1}(t)x + S_{\alpha,2}(t)y + \int_0^t S_{\alpha,\alpha}(t-s)f(s)ds, \quad t \geq 0.$$

As in the proof of Proposition A.1, we have that the convolution $S_{\alpha,\alpha} * f$ belongs to $C_0(\mathbb{R}_+, X)$, and by Corollary 4.1(a),

$$\|u(t)\| \leq \|S_{\alpha,1}(t)\| \, \|x\| + \|S_{\alpha,2}(t)\| \, \|y\| + \|(S_{\alpha,\alpha} * f)(t)\| \to 0,$$

as $t \to \infty$. Therefore, $u \in C_0(\mathbb{R}_+, X)$. $\qquad\square$

Theorem A.12 (Ponce, 2020b, Theorem 4.12). *Let $1 < p < \infty, 1 < \alpha < 2$, and $\beta \geq 1$ such that $p(\alpha - \beta + 1) > 1$. If A is ϖ-sectorial of angle $\theta = \frac{(\alpha-1)}{2}\pi$, with $\varpi < 0$, then $\left\{S_{\alpha,\beta}(t)\right\}_{t \geq 0}$ is uniformly p-integrable. Particularly, $\left\{S_{\alpha,\alpha}(t)\right\}_{t \geq 0}$ is uniformly p-integrable.*

Proof. By Theorem 4.3, there exists a constant $C > 0$ such that $\|S_{\alpha,\beta}(t)\| \leq \frac{Ct^{\beta-1}}{(1+|\varpi|t^\alpha)}$ for all $t \geq 0$. It follows from the assumptions on α, β and p that

$$\int_0^\infty \|S_{\alpha,\beta}(t)\|^p dt \leq \int_0^\infty \frac{C^p t^{(\beta-1)p}}{(1+|\varpi|t^\alpha)^p} dt$$

$$= \frac{C^p}{\alpha} \frac{1}{|\varpi|^{(\beta-1)p/\alpha+1/\alpha-1}} \mathbf{B}\left(\frac{(\beta-1)p}{\alpha} + \frac{1}{\alpha}, p\left(1 - \frac{\beta}{\alpha}\right) + \frac{1}{\alpha}(p-1)\right)$$

$$< \infty.$$

This proves the assertion. $\qquad\square$

In the following results, we obtain L^p-regularity of the solutions to the problem (A.13).

Corollary A.3 (Ponce, 2020b, Corollary 4.13). *Let $1 < p < \infty$, $1 < \alpha < 2$, $\varpi < 0$ and A is ϖ-sectorial of angle $\theta = \frac{(\alpha-1)}{2}\pi$ defined on a Banach space $(X, \|\cdot\|)$. If $f \in L^q(\mathbb{R}_+, X)$, $1/p + 1/q = 1$, then the solution u to the problem (A.13) verifies $\|u(t)\| \to 0$, as $t \to \infty$.*

Proof. Since $\{S_{\alpha,\alpha}(t)\}_{t\geq 0}$ is uniformly p-integrable by Theorem A.12 and $f \in L^q(\mathbb{R}_+, X)$, we obtain that $S_{\alpha,\alpha} * f \in C_0(\mathbb{R}_+, X)$ (see Lemma A.3(b) or Arendt *et al.*, 2001, Proposition 1.3.5). Because the solution to the problem (A.13) is given by (A.14), the Corollary 4.1(a) implies that $u \in C_0(\mathbb{R}_+, X)$. $\square$

Corollary A.4 (Ponce, 2020b, Corollary 4.14). *Let $1 < p < \infty$, $1 < \alpha < 2$, $\varpi < 0$ and A is ϖ-sectorial of angle $\theta = \frac{(\alpha-1)}{2}\pi$ defined on a Banach space $(X, \|\cdot\|)$. If $f \in L^1(\mathbb{R}_+, X)$ then the solution u to the problem (A.13) belongs to $L^p(\mathbb{R}_+, X)$.*

Proof. By Young's inequality (see Lemma A.3(a) or Arendt *et al.* (2001, Proposition 1.3.5)) and Theorem A.12, it follows that

$$\|S_{\alpha,\alpha} * f\|_p \leq \|f\|_1 \left(\int_0^\infty \|S_{\alpha,\alpha}(t)\|^p dt \right)^{1/p} < \infty,$$

i.e., $S_{\alpha,\alpha} * f \in L^p(\mathbb{R}_+, X)$. Since the solution to the problem (A.13) is given by (A.14), Theorem A.12 implies that $u \in L^p(\mathbb{R}_+, X)$. $\square$

Similar to the previous results, for the Caputo fractional Cauchy problem (A.9) we have the following corollaries.

Corollary A.5 (Ponce, 2020b, Corollary 4.15). *Let $1 < p < \infty$, $1 < \alpha < 2$, $\varpi < 0$ and A is ϖ-sectorial of angle $\theta = \frac{(\alpha-1)}{2}\pi$ defined on a Banach space $(X, \|\cdot\|)$. If $f \in L^q(\mathbb{R}_+, X)$, $1/p+1/q = 1$, then the solution u to the problem (A.9) verifies $\|u(t)\| \to 0$, as $t \to \infty$.*

Corollary A.6 (Ponce, 2020b, Corollary 4.16). *Let $1 < p < \infty$, $1 < \alpha < 2$, $\varpi < 0$ and A is ϖ-sectorial of angle $\theta = \frac{(\alpha-1)}{2}\pi$ defined on a Banach space $(X, \|\cdot\|)$. If $f \in L^1(\mathbb{R}_+, X)$ and $p(\alpha - 1) > 1$, then the solution u to the problem (A.9) belongs to $L^p(\mathbb{R}_+, X)$.*

Proof. As in the proof of Corollary A.4, $S_{\alpha,\alpha} * f \in L^p(\mathbb{R}_+, X)$. The assumption $p(\alpha - 1) > 1$ implies that $S_{\alpha,1}$ and $S_{\alpha,2}$ belong to $L^p(\mathbb{R}_+, X)$ by Theorem A.12. Therefore, $u \in L^p(\mathbb{R}_+, X)$. $\square$

For the problem (A.11). Since $1 < \alpha < 2$, it is noted that Theorem 4.3 of Chapter 4 cannot ensure that $\|S_{\alpha,\alpha-1}(t)\| \to 0$ as $t \to \infty$. However, the following result is true, which implies that in order to guarantee the convergence to zero of the solution u to the problem (A.11), we need to integrate $(\alpha - 1)$-times the solution u.

Proposition A.3 (Ponce, 2020b, Corollary 4.17). *Let* $3/2 < \alpha < 2$, $\varpi < 0$ *and* A *is* ϖ-*sectorial of angle* $\theta = \frac{(\alpha-1)}{2}\pi$ *defined on a Banach space* $(X, \|\cdot\|)$. *If* $f \in L^1(\mathbb{R}_+, X)$, *then the mild solution* u *to the problem (A.11) satisfies* $\|(g_{\alpha-1} * u)(t)\| \to 0$ *as* $t \to \infty$.

Proof. It is observed that for all $\mathrm{Re}\lambda > 0$,

$$\big(\widehat{g_{\alpha-1} * S_{\alpha,\alpha-1}}\big)(\lambda) = \frac{1}{\lambda^{\alpha-1}}\lambda\big(\lambda^\alpha - A\big)^{-1}$$

$$= \lambda^{\alpha-(2\alpha-2)}\big(\lambda^\alpha - A\big)^{-1} = \big(\widehat{S_{\alpha,2\alpha-2}}\big)(\lambda).$$

Then it is concluded by the uniqueness of the Laplace transform that $\big(g_{\alpha-1}* S_{\alpha,\alpha-1}\big)(t) = S_{\alpha,2\alpha-2}(t)$. Since $3/2 < \alpha < 2$, we can apply Corollary 4.1(a) of Chapter 4 to obtain that $\|(g_{\alpha-1}*S_{\alpha,\alpha-1})(t)\| \to 0$ as $t \to \infty$. Analogously, $\big(\widehat{g_{\alpha-1} * S_{\alpha,\alpha}}\big)(\lambda) = \lambda^{\alpha-(2\alpha-1)}(\lambda^\alpha - A)^{-1}$ and therefore $\big(g_{\alpha-1} * S_{\alpha,\alpha}\big)(t) = S_{\alpha,2\alpha-1}(t)$. It follows again by Corollary 4.1(a) of Chapter 4 that $\big(g_{\alpha-1} * S_{\alpha,\alpha}\big)(t) \to 0$ as $t \to \infty$. Finally, the convolution $g_{\alpha-1} * S_{\alpha,\alpha} * f$ belongs to $C_0(\mathbb{R}_+, X)$ by Lemma A.3(d) (or Arendt *et al.*, 2001, Proposition 1.3.5), and by (A.12), we conclude that $g_{\alpha-1} * u$ approaches 0 as $t \to \infty$. □

Remark A.4. According to Theorems 5.2, 5.3, of Chapter 5 and Theorem A.9, we can establish some existence results of generalized Bloch-type periodic mild solution to Eq. (5.1) of Chapter 5 when f is not necessarily Lipschitz with its second variable, which can be accomplished similarly to Theorem 4.8 of Chapter 4.

Bibliography

Abadias, L. and Miana, P. J. (2015). A subordination principle on Wright functions and regularized resolvent families, *J. Function Spaces*, **2015**, Article ID 158145, pp. 9.

Abbas, S., Benchohra, M. and N'Guérékata, G. M. (2012). *Topics in Fractional Differential Equations* (Springer, New York).

Agrawal, O., Sabatier, J. and Tenreiro, J. (2007). *Advances in Fractional Calculus* (Springer, Dordrecht).

Al-Islam, N. S., Alsulami, S. M. and Diagana, T. (2012). Existence of weighted pseudo anti-periodic solutions to some non-autonomous differential equations, *Appl. Math. Comput.* **218**, pp. 6536–6648.

Alvarez, E., Gomez, A. and Pinto, M. (2018). (ω, c)-periodic functions and mild solution to abstract fractional integro-differential equations, *Electron. J. Qual. Theory Differ. Equ.* **2018**, 8 pp.

Alvarez, E., Lizama, C. and Ponce, R. (2015). Weighted pseudo antiperiodic solutions for fractional integro-differential equations in Banach spaces, *Appl. Math. Comput.* **259**, pp. 164–172.

Alvarez-Pardo, E. and Lizama, C. (2015). Weighted pseudo almost automorphic mild solutions for two-term fractional order differential equations, *Appl. Math. Comput.* **271**, pp. 154–167.

Andrade, B. de and Cuevas, C. (2010). S-asymptotically ω-periodic and asymptotically ω-periodic solutions to semi-linear Cauchy problems with non-dense domain, *Nonlin. Anal.* **72**, pp. 3190–3208.

Andrade, B. de, Cuevas, C. and Henríquez, E. (2012). Asymptotic periodicity and almost automorphy for a class of Volterra integro-differential equations, *Math. Meth. Appl. Sci.* **35**, pp. 795–811.

Andrade, B. de, Cuevas, C., Silva, C. and Soto, H. (2016). Asymptotic periodicity for flexible structural systems and applications, *Acta Appl. Math.* **143**, pp. 105–164.

Andrade, F., Cuevas, C. and Henríquez, H. R. (2021). Existence of asymptotically periodic solutions of partial functional differential equations with state-dependent delay, *Appl. Anal.* **100**, pp. 2965–2988.

Andrade, B. de and Lizama, C. (2011). Existence of asymptotically almost periodic solutions for damped wave equations, *J. Math. Anal. Appl.* **382**, pp. 761–771.

Araya, D. and Lizama, C. (2008). Almost automorphic mild solutions to fractional differential equations, *Nonlin. Anal.* **69**, pp. 3692–3705.

Arendt, W. and Favini, A. (1993). Integrated solutions to implicit differential equations, *Rend. Sem. Mat. Univ. Pol. Torino* **51**, pp. 315–329.

Arendt, W., Batty, C., Hieber, M. and Neubrander, F. (2001). *Vector-Valued Laplace Transforms and Cauchy Problems* (Birkhäuser, Basel).

Bachir, F. S., Abbas, S., Benbachir, M., Benchohra, M., and N'Guérékata, G. M. (2021). Existence and attractivity results for ψ-Hilfer hybrid fractional differential equations, *Cubo* **23**, pp. 145–159.

Barbu, V. and Favini, A. (1997). Periodic problems for degenerate differential equations, *Rend. Instit. Mat. Univ. Trieste* **XXVIII** (Suppl.), pp. 29–57.

Bedi, P., Kumar, A., Abdeljawad, T. and Khan, A. (2021). S-asymptotically ω-periodic mild solutions and stability analysis of Hilfer fractional evolution equations, *Evol. Equ. Control Theo.* **10**, pp. 733–748.

Benedetti, I., Obukhovskii, V. and Taddei, V. (2020). Evolution fractional differential problems with impulses and nonlocal conditions, *Discrete Contin. Dyn. Syst.-S* **13**, pp. 1899–1919.

Bian, Y. T., Chang, Y. K. and Nieto, J. J. (2014). Weighted asymptotic behavior of solutions to a semilinear integro-differential equation in Banach spaces, *Electron. J. Diff. Equ.* **2014**(91), pp. 1–16.

Bloch, F. (1929). Über die quantenmechanik der elektronen in kristallgittern, *Z. Physik* **52**, 555–600.

Blot, J., Cieutat, P. and Ezzinbi, K. (2012). Measure theory and pseudo almost automorphic functions: New development and applications, *Nonlin. Anal.* **75**, pp. 2426–2447.

Bose, S. K. and Gorain, G. C. (1998a). Stability of the boundary stablisized damped wave equation $y'' + \lambda y''' = c^2(\Delta y + \mu \Delta y')$ in a bounded domain in $\mathbb{R}^n$, *Indian J. Math.* **40**, pp. 1–15.

Bose, S. K. and Gorain, G. C. (1998b). Exact controllability and boundary stablization of torsional virations of an internally damped flexible space structure, *J. Optim. Theory Appl.* **99**, 423–442.

Brindle, D. and N'Guérékata, G. M. (2019a). *S-asymptotically ω periodic functions and sequences and applications to evolution equations*, Ph.D. thesis, Morgan State University.

Brindle, D. and N'Guérékata, G. M. (2019b). S-asymptotically ω-periodic sequential equations to difference equations, *Nonlin. Stud.* **26**, pp. 575–586.

Brindle, D. and N'Guérékata, G. M. (2019c). Existence results of S-asymptotically τ-periodic mild solutions to integrodifferential equations, *PanAmer. Math. J.* **29**, pp. 63–74.

Brindle, D. and N'Guérékata, G. M. (2020). S-asymptotically ω-periodic mild solutions to fractional differential equations, *Electron. J. Diff. Equ.* **2020**(30), pp. 1–12.

Burlică, M. D., Necula, M., Daniela, R. and Vrabie, I. I. (2016). *Delay Differential Evolutions Subjected to Nonlocal Initial Conditions* (Chapman & Hall/CRC, Boca Raton).

Byszewski, L. and Lakshmikantham, V. (1991). Theorem about the existence and uniqueness of a solutions of a nonlocal abstract Cauchy problem in a Banach space, *Appl. Anal.* **40**, pp. 11–19.

Cao, J., Yang, Q. and Huang, Z. (2012). Existence of anti-periodic mild solutions for a class of semilinear fractional differential equations, *Commun. Nonlin. Sci. Numer. Simulat.* **17**, pp. 277–283.

Cao, J. and Huang, Z. (2018). Existence of asymptotically periodic solutions for semilinear evolution equations with nonlocal initial conditions, *Open Math.* **16**, pp. 792–805.

Carroll, R. W. and Showalter, R. E. (1976). *Singular and Degenerate Cauchy Problems* (Academic Press, New York).

Chang, Y. K. and Li, W. T. (2006). Existence results for impulsive dynamic equations on time scales with nonlocal initial conditions, *Math. Comput. Model.* **43**, pp. 377–384.

Chang, Y. K. and Luo, X. X. (2015). Pseudo almost automorphic behavior of solutions to a semi-linear fractional differential equation, *Math. Commun.* **20**, pp. 53–68.

Chang, Y. K. and Pei, Y. (2020). Degenerate type fractional evolution hemivariational inequalities and optimal controls via fractional resolvent operators, *Int. J. Control* **93**, pp. 528–540.

Chang, Y. K., Luo, X. X. and N'Guérékata, G. M. (2014). Asymptotically typed solutions to a semilinear integral equation, *J. Integral Equ. Appl.* **26**, pp. 323–343.

Chang, Y. K., N'Guérékata, G. M. and Zhao, Z. H. (2018). Recurrence of bounded solutions to a semilinear integro-differential equation perturbed by Lévy noise, *Asymptot. Anal.* **109**, pp. 27–52.

Chang, Y. K., Nieto, J. J. and Li, W. S. (2009). On Impulsive Hyperbolic Differential Inclusions with Nonlocal Initial Conditions, *J. Optim. Theory Appl.* **140**, pp. 431–442.

Chang, Y. K., Pei, Y. and Ponce, R. (2019a). Existence and optimal controls for fractional stochastic evolution equations of Sobolev type via fractional resolvent operators, *J. Optim. Theory Appl.* **182**, pp. 558–572.

Chang, Y. K. and Ponce, R. (2018). Uniform exponential stability and its applications to bounded solutions of integro-differential equations in Banach spaces, *J. Integral Equ. Appl.* **30**, pp. 347–369.

Chang, Y. K. and Ponce, R. (2019). Sobolev type time fractional differential equations and optimal controls with the order in $(1, 2)$, *Differ. Integral Equ.* **32**, pp. 517–540.

Chang, Y. K. and Ponce, R. (2021). Mild solutions for a multi-term fractional differential equation via resolvent operators, *AIMS Math.* **6**, pp. 2398–2417.

Chang, Y. K. and Wei, Y. (2021a). S-asymptotically Bloch type periodic solutions to some semi-linear evolution equations in Banach spaces, *Acta Math. Sci.* **41**, pp. 413–425.

Chang, Y. K. and Wei, Y. (2021b). Pseudo S-asymptotically Bloch type periodicity with applications to some evolution equations, *Z. Anal. Anwend.* **40**, pp. 33–50.

Chang, Y. K. and Wei, Y. (2022). Pseudo S-asymptotically Bloch type periodic solutions to fractional integro-differential equations with Stepanov-like force terms, *Z. Angew. Math. Phys.* **73**, Art. 77, 17 pp.

Chang, Y. K., Pereira, A. and Ponce, R. (2017). Approximate controllability for fractional differential equations of Sobolev type via properties on resolvent operators, *Fract. Calc. Appl. Anal.* **20**, pp. 963–987.

Chang, Y. K., Ponce, R. and Rueda, S. (2019b). Fractional differential equations of Sobolev type with sectorial operators, *Semigroup Forum* **99**, pp. 591–606.

Chang, Y. K., Ponce, R. and Yang, X. S. (2021). Solvability of fractional differential inclusions with nonlocal initial conditions via resolvent family of operators, *Int. J. Nonlinear Sci. Numer. Simul.* **22**, pp. 33–44.

Chang, Y. K., Wei, X. Y. and N'Guérékata, G. M. (2015b). Some new results on bounded solutions to a semilinear integro-differential equation in Banach spaces, *J. Integral Equ. Appl.* **27**, pp. 153–178.

Chang, Y. K., Zhang, R. and N'Guérékata, G. M. (2012). Weighted pseudo almost automorphic mild solutions to semilinear fractional differential equations, *Comput. Math. Appl.* **64**, pp. 3160–3170.

Chang, Y. K., Zhang, M. J. and Ponce, R. (2015a). Weighted pseudo almost automorphic solutions to a semilinear fractional differential equation with Stepanov-like weighted pseudo almost automorphic nonlinear term, *Appl. Math. Comput.* **257**, pp. 158–168.

Chaouchi, B., Kostić, M., Pilipovic, S., and Velinov, D. (2020). Semi-Bloch periodic functions, semi-anti-periodic functions and applications, *Chel. Phy. Math. J.* **5**, pp. 243–255.

Chen, D. H. and Wang, R. N. (2015). New qualitative properties of solutions to nonlinear nonlocal Cauchy problems, *Rocky Mount. J. Math.* **45**, pp. 427–456.

Chen, S., Chang, Y. K. and Wei, Y. (2022). Pseudo S-asymptotically Bloch type periodic solutions to a damped evolution equation, *Evol. Equ. Control Theo.* **11**, pp. 621–633.

Chen, P., Li, Y. and Zhang, X. (2018). *Solvability of Abstract Evolution Equations with Nonlocal Conditions and Applications* (Science Press, Beijing).

Chen, J. H., Xiao, T. J. and Liang, J. (2009). Uniform exponential stability of solutions to abstract Volterra equations, *J. Evol. Equ.* **4**, pp. 661–674.

Clément, Ph. and Prato, G. Da (1998). Existence and regularity results for an integral equation with infinite delay in a Banach space, *Integr. Eqn. Oper. Theory* **11**, pp. 480–500.

Cuesta, E. (2007). Asymptotic behaviour of the solutions of fractional integro-differential equations and some time discretizations, *Discrete Contin. Dyn. Syst.* **suppl.**, pp. 277–285.

Cuesta, E. and Ponce, R. (2018). Well-posedness, regularity, and asymptotic behavior of the continuous and discrete solutions of linear fractional integro-differential equations with order varying in time, *Electron. J. Diff. Equ.* **2018**(173), pp. 1–27.

Cuevas, C. and Henríquez, H. R. (2011). Solutions of second order abstract retarded functional differential equations on the line, *J. Nonlinear Convex Anal.* **12**, pp. 225–240.

Cuevas, C. and Lizama, C. (2008). Almost automorphic solutions to a class of semilinear fractional differential equations, *Appl. Math. Lett.* **21**, pp. 1315–1319.

Cuevas, C. and Lizama, C. (2009). Almost automorphic solutions to integral equations on the line, *Semigroup Forum* **79**, pp. 461–472.

Cuevas, C. and Lizama, C. (2010). *S*-asymptotically ω-periodic solutions for semilinear Volterra equations, *Math. Meth. Appl. Sci.* **33**, pp. 1628–1636.

Cuevas, C. and Souza, J. C. de (2009). *S*-asymptotically ω-periodic solutions of semilinear fractional integro-differential equations, *Appl. Math. Lett.* **22**, pp. 865–870.

Cuevas, C. and Souza, J. C. de (2010). Existence of *S*-asymptotically ω-periodic solutions for fractional order functional integro-differential equations with infinite delay, *Nonlinear Anal.* **72**, pp. 1683–1689.

Cuevas, C., Henríquez, H. R. and Soto, H. (2014). Asymptotically periodic solutions of fractional differential equations, *Appl. Math. Comput.* **236**, pp. 524–545.

Cui, N. and Sun, H. R. (2021a). Existence and multiplicity results for the fractional Schrödinger equations with indefinite potentials, *Appl. Anal.* **100**, pp. 1198–1212.

Cui, N. and Sun, H. R. (2021b). Fractional p-Laplacian problem with indefinite weight in R^N: Eigenvalues and existence, *Math. Meth. Appl. Sci.* **44**, pp. 2585–2599.

Davies, E. and Pang, M. (1987). The Cauchy problem and a generalization of the Hille-Yosida theorem, *Proc. London Math. Soc.* **55**, pp. 181–208.

Debbouche, A. and Nieto, J. J. (2014). Sobolev type fractional abstract evolution equations with nonlocal conditions and optimal multi-controls, *Appl. Math. Comput.* **245**, pp. 74–85.

Debbouche, A. and Torres, D. (2015). Sobolev type fractional dynamic equations and optimal multi-integral controls with fractional nonlocal conditions, *Fract. Calc. Appl. Anal.* **18**, pp. 95–121.

Diestel, J. and Uhl, JR., J. J. (1977). *Vector Measures* (Am. Math. Soc., Providence, Rhode Island).

Dimbour, W. (2020). Pseudo *S*-asymptotically ω-periodic solution for a differential equation with piecewise constant argument in a Banach space, *J. Differ. Equ. Appl.* **26**, pp. 140–148.

Ding, H. S., Xiao, T. J. and Liang, J. (2008). Asymptotically almost automorphic solutions for some integrodifferenial equations with nonlocal initial conditions, *J. Math. Anal. Appl.* **338**, pp. 141–151.

Engel, K. J. and Nagel, R. (1999). *One-Parameter Semigroups for Linear Evolution Equations* (Springer, New York).

Fan, Z. (2014). Characterization of compactness for resolvents and its applications, *Appl. Math. Comput.* **232**, pp. 60–67.

Fan, Z. and Li, G. (2010). Existence results for semilinear differential equations with nonlocal and impulsive conditions, *J. Funt. Anal.* **258**, pp. 1709–1727.

Favini, A. and Yagi, A. (1999). *Degenerate Differential Equations in Banach Spaces* (Pure and Applied Math., **215**, Dekker, New York).

Fu, X. and Zhang, Y. (2013). Exact null controllability of non-autonomous functional evolution systems with nonlocal conditions, *Acta Math. Sci.*, **33**, pp. 747–757.

Gao, H., Wang, K., Wei, F. and Ding, X. (2006). Massera-type theorem and asymptotically periodic Logisitc equations, *Nonlin. Anal. RWA* **7**, pp. 1268–1283.

Gorain, G. C. (2006). Boundary stablization of nonlinear vibrations of a flexible structure in a bounded domain in $\mathbb{R}^N$, *J. Math. Anal. Appl.* **319**, pp. 635–650.

Gorenflo, R. and Mainardi, F. (1997). Fractional calculus: Integral and differential equations of fractional order, in: A. Carpinteri, F. Mainardi (Eds.), *Fractals and Fractional Calculus in Continoum Mechanics* (Springer, New York), pp. 223–276.

Granas, A. and Dugundji, J. (2003). *Fixed Point Theory* (Springer, New York).

Haase, M. (2006). *The Functional Calculus for Sectorial Operators* (Birkäuser Verlag, Basel).

Haase, M. (2008). The complex inversion formula revisited, *J. Aust. Math. Soc.* **84**, pp. 73–83.

Hasler, M. F. and N'Guérékata, G. M. (2014). Bloch-periodic functions and some applications, *Nonlin. Stud.* **21**, pp. 21–30.

He, B., Wang, Q. R. and Cao J. F. (2020). Weighted S^p-pseudo S-asymptotic periodicity and applications to Volterra integral equations, *Appl. Math. Comput.* **380**, 125275.

Henríquez, H. R. (2013). Asymptotically periodic solutions of abstract differential equations, *Nonlin. Anal.* **80**, pp. 135–149.

Henríquez, H. R. and Lizama, C. (2009). Compact almost automorphic solutions to integral equations with infinite delay, *Nonlin. Anal.* **71**, pp. 6029–6037.

Henríquez, H. R., Cuevas, C. and Caicedo, A. (2013). Asymptotically periodic solutions of neutral partial differential equations with infinite delay, *Commu. Pure Appl. Anal.* **12**, pp. 2031–2068.

Henríquez, H. R., Pierri, M. and Rolnik, V. (2016). Pseudo S-asymptotically periodic solutions of second-order abstract Cauchy problems, *Appl. Math. Comput.* **274**, pp. 590–603.

Henríquez, H. R., Pierri, M. and Táboas, P. (2008a). On S-asymptotically ω-periodic functions on Banach spaces and applications, *J. Math. Anal. Appl.* **343**, pp. 1119–1130.

Henríquez, H. R., Pierri, M. and Táboas, P. (2008b). Existence of S-asymptotically ω-periodic solutions for abstract neutral functional-differential equations, *Bull. Aust. Math. Soc.* **78**, pp. 365–382.

Hernández, M. E. and Dos Santos, J. P. C. (2016). Asymptotically almost periodic and almost periodic solutions for a class of partial integrodifferential equations, *Electron. J. Diff. Equ.* **2006**(38), pp. 1–8.

Hernández, E. and O'Regan, D. (2018). On state dependent non-local conditions, *Appl. Math. Lett.* **83**, pp. 103–109.

Hernández, E. and Pierri, M. (2018). *S*-asymptotically periodic solutions for abstract equations with state-dependent delay, *Bull. Aust. Math. Soc.* **98**, pp. 456–464.

Hilfer, R. (ed.) (2000). *Applications of Fractional Calculus in Physics* (World Scientific, River Edge, NJ).

Keyantuo, V., Lizama, C. and Warma, M. (2013). Asymptotic behavior of fractional-order semilinear evolution equation, *Differ. Integral Equ.* **26**, pp. 757–780.

Khalladi, M. T., Kostic, M., Rahmani, A., Pinto, M. and Velinov, D. (2021). On semi-*c*-periodic functions, *J. Math.* (Hindawi) **2021**, Article ID 6620625, 5 pp.

Kilbas, A., Srivastava, H. and Trujillo, J. (2006). *Theory and Applications of Fractional Differential Equations* (Elsevier, Amsterdam).

Kittel, C. (2005). *Introduction to Solid State Physics* 8th ed., (John Wiley & Sons, New York).

Kostić, M. (2019). *Almost Periodic and Almost Automorphic Solutions to Integro-Differential Equations* (W. de Gruyter, Berlin).

Kostić, M. and Velinov, D. (2017). Asymptotically Bloch-periodic solutions of abstract fractional nonlinear differential inclusions with piecewise constant argument, *Funct. Anal. Approx. Comput.* **9**, pp. 27–36.

Lang, S. (1999). *Complex Analysis* (Springer, Berlin).

Li, F., Liang, J. and Xu, H. K. (2012). Existence of mild solutions for fractional integrodifferential equations of Sobolev type with nonlocal conditions, *J. Math. Anal. Appl.* **391**, pp. 510–525.

Liang, J., Mu, Y. and Xiao, T. J. (2019). Solutions to fractional Sobolev-type integro-differential equations in Banach spaces with operator pair and impulsive conditions, *Banach J. Math. Anal.* **13**, pp. 745–768.

Liu, Z. (2010). Anti-periodic solutions to nonlinear evolution equations, *J. Func. Anal.* **258**, pp. 2026–2033.

Liu, Z. and Li, X. (2015). Approximate controllability of fractional evolution systems with Riemann–Liouville fractional derivative, *SIAM J. Control Optim.* **53**, pp. 1920–1933.

Lizama, C. (2000). Regularized solutions for abstract Volterra equations, *J. Math. Anal. Appl.* **243**, pp. 278–292.

Lizama, C. and N'Guérékata, G. M. (2010). Bounded mild solutions for semilinear integro-differential equations in Banach spaces, *Integr. Eqn. Oper. Theory* **68**, pp. 207–227.

Lizama, C. and Poblete, F. (2012). On a functional equation associated with (a, k)-regularized resolvent families, *Abstr. Appl. Anal*, vol. 2012, Article ID 495487, 23 pp.

Lizama, C. and Poblete, V. (2007). On multiplicative perturbation of integral resolvent families, *J. Math. Anal. Appl.* **327**, pp. 1335–1359.

Lizama, C. and Ponce, R. (2011a). Periodic solutions of degenerate differential equations in vector-valued function spaces, *Studia Math.* **202**, pp. 49–63.

Lizama, C. and Ponce, R. (2011b). Bounded solutions to a class of semilinear integrodifferential equations in Banach spaces, *Nonlin. Anal.* **74**, pp. 3397–3406.

Lizama, C. and Ponce, R. (2011c). Almost automorphic solutions to abstract Volterra equations on the line, *Nonlin. Anal.* **74**, pp. 3805–3814.

Lizama, C. and Rueda, S. (2019). Nonlocal integrated solutions for a class of abstract evolution equations, *Acta Appl. Math.* **164**, pp. 165–183.

Lizama, C., Pereira, A. and Ponce, R. (2016). On the compactness of fractional resolvent operator functions, *Semigroup Forum* **93**, pp. 363–374.

Lü, Q. and Zuazua, E. (2016). On the lack of controllability of fractional in time ODE and PDE, *Math. Control Sign. Syst.* **28**, Art. 10, 21 pp.

Mainardi, F. (2010). *Fractional Calculus and Waves in Linear Viscoelasticity: An Introduction to Mathematical Models* (Imperial College Press, London).

Mátrai, T. (2008). Resolvent norm decay does not characterize norm continuity, *Israel J. Math.* **168**, pp. 1–28.

Melnikova, I. V. and Filinkov, A. (2001). *Abstract Cauchy Problems: Three Approaches* (Chapman & Hall/CRC, Boca Raton).

Meyers, M. A. and Chawla, K. K. (2009). *Mechanical Behavior of Materials* (Cambridge University Press, Cambridge).

Mophou, G. M. (2011). Weighted pseudo almost automorphic mild solutions to semilinear fractional differential equations, *Appl. Math. Comput.* **217**, pp. 7579–7587.

Mophou, G. M. and N'Guérékata, G. M. (2020). An existence result of (ω, c)-periodic mild solutions to some fractional differential equation, *Nonlin. Stud.* **27**, pp. 167–175.

N'Guérékata, G. M. (2005). *Topics in Almost Automorphy* (Springer, New York).

N'Guérékata, G. M. (2021). *Almost Periodic and Almost Automorphic Functions in Abstract Spaces* (Springer, New York).

Oueama-Guengai, E. R. (2018). S-asymptotically ω-periodic mild solutions to some fractional integrodifferential equations with infinite delay, *Lib. Math (NS)* **38**, pp. 111–124.

Oueama-Guengai, E. R. and N'Guérékata, G. M. (2018). On S-asymptotically ω-periodic and Bloch periodic mild solutions to some fractional differential equations in abstract spaces, *Math. Meth. Appl. Sci.* **41**, pp. 9116–9122.

Pazy, A. (1983). *Semigroups of Linear Operators and Applications to Partial Differential Equations* (Springer, New York).

Pei, Y. and Chang, Y. K. (2020). Approximate controllability for stochastic fractional hemivariational inequalities of degenerate type, *Fract. Calc. Appl. Anal.* **23**, pp. 1506–1531.

Pierri, M. (2012). On S-asymptotically ω-periodic functions and applications, *Nonlinear Anal.* **75**, pp. 651–661.

Pierri, M. and Nicola, S. (2009). A note on S-asymptotically periodic functions, *Nonlinear Anal. RWA* **10**, pp. 2937–2938.

Pierri, M. and Rolnik, V. (2013). On pseudo S-asymptotically periodic functions, *Bull. Aust. Math. Soc.* **87**, pp. 238–254.

Ponce, R. (2013a). Bounded mild solutions to fractional integro-differential equations in Banach spaces, *Semigroup Forum* **87**, pp. 377–392.

Ponce, R. (2013b). Hölder continuous solutions for fractional differential equations and maximal regularity, *J. Diff. Equ.* **10**, pp. 3284–3304.

Ponce, R. (2014). Hölder continuous solutions for Sobolev type differential equations, *Math. Nachr.* **287**, pp. 70–78.

Ponce, R. (2016). Existence of mild solutions to nonlocal fractional Cauchy problems via compactness, *Abstr. Appl. Anal.* Art. ID 4567092, 15 pp.

Ponce, R. (2017). On the well-posedness of degenerate fractional differential equations in vector valued function spaces, *Israel J. Math.* **219**, pp. 727–755.

Ponce, R. (2020a). Mild solutions to integro-differential equations in Banach spaces, *J. Diff. Equ.* **269**, pp. 180–200.

Ponce, R. (2020b). Asymptotic behavior of mild solutions to fractional Cauchy problems in Banach spaces, *Appl. Math. Lett.* **105**, 106322.

Ponce, R. (2020c). Time discretization of fractional subdiffusion equations via fractional resolvent operators, *Comput. Math. Appl.* **80**, pp. 69–92.

Ponce, R. (2020d). Subordination principle for fractional diffusion of Sobolev type, *Fract. Calc. Appl. Anal.* **23**, pp. 427–449.

Prüss, J. (1993). *Evolutionary Integral Equations and Applications* (Birkhäuser Verlag).

Prüss, J. (2009). Decay properties for the solutions of a partial differential equation with memory, *Arch. Math. (Basel)* **92**, pp. 158–173.

Ren, S. Y. (2006). *Electronic States in Crystals of Finite Size: Quantum Confinement of Bloch Waves* (Springer, New York).

Shu, X. B., Xu, F. and Shi, Y. (2015). S-asymptotically ω-positive periodic solutions for a class of neutral fractional differential equations, *Appl. Math. Comput.* **270**, pp. 768–776.

Sviridyuk, G. and Fedorov, V. (2003). *Linear Sobolev Type Equations and Degenerate Semigroups of Operators* (De Gruyter, Berlin).

Wang, G. (2018). Twin iterative positive solutions of fractional q-difference Schrödinger equations, *Appl. Math. Lett.* **76**, pp. 103–109.

Wang, R. N., Chen, D. H. and Xiao, T. J. (2012). Abstract fractional Cauchy problems with almost sectorial operators, *J. Diff. Equ.* **252**, pp. 202–235.

Wang, G., Ren, X., Bai, Z. and Hou, W. (2019). Radial symmetry of standing waves for nonlinear fractional Hardy-Schrödinger equation, *Appl. Math. Lett.* **96**, pp. 131–137.

Wei, Y. and Chang, Y. K. (2022). Generalized Bloch type periodicity and applications to semi-linear differential equations in Banach spaces, *Proc. Edinburgh Math. Soc.* https://dx.doi.org/10.1017/S0013091522000098

Weis, L. W. (1988). A generalization of the Vidav-Jorgens perturbation theorem for semigroups and its application to transport theory, *J. Math. Anal. Appl.* **129**, pp. 6–23.

Xia, Z. (2014). Weighted Stepanov-like pseudoperiodicity and applications, *Abstr. Appl. Anal.* Art. ID 980868, 14 pp.

Xia, Z. (2015a). Pseudo asymptotically periodic solutions for Volterra integro-differential equations, *Math. Meth. Appl. Sci.* **38**, pp. 799–810.

Xia, Z. (2015b). Weighted pseudo asymptotically periodic mild solutions of evolution equations, *Acta Math. Sinica* **31**, pp. 1215–1232.

Xia, Z., Wang, D., Wen, C. F. and Yao, J. C. (2017). Pseudo asymptotically periodic mild solutions of semilinear functional integro-differential equations in Banach spaces, *Math. Meth. Appl. Sci.* **40**, pp. 7333–7355.

Yan, Z. and Jia, X. (2015). Approximate controllability of partial fractional neutral stochastic functional integro-differential inclusions with state-dependent delay, *Collect. Math.* **66**, pp. 93–124.

Yan, Z. and Jia, X. (2017). Optimal controls of fractional impulsive partial neutral stochastic integro-differential systems with infinite delay in Hilbert spaces, *Int. J. Control Autom. Syst.* **15**, pp. 1051–1068.

Yang, M. and Wang, Q. R. (2019). Pseudo asymptotically periodic solutions for fractional integro-differential neutral equations, *Sci. China Math.* **62**, pp. 1705–1718.

You, P. (1992). Characteristic conditions for a C_0-semigroup with continuity in the uniform operator topology for $t > 0$ in Hilbert space, *Proc. Amer. Math. Soc.* **116**, pp. 991–997.

Zhang, R., Chang, Y. K. and N'Guérékata, G. M. (2013). Weighted pseudo almost automorphic mild solutions to semilinear integral equations with S^p-weighted pseudo almost automorphic coefficients, *Discrete Contin. Dyn. Syst.* **33**, pp. 5525–5537.

Zhao, J. Q., Chang, Y. K. and N'Guérékata, G. M. (2013a). Asymptotic behavior of mild solutions to semilinear fractional differential equations, *J. Optim. Theory Appl.* **156**, pp. 106–114.

Zhao, Z. H. and Chang, Y. K. (2020). Topological properties of solution sets for Sobolev-type fractional stochastic differential inclusions with Poisson jumps, *Appl. Anal.* **99**, pp. 1373–1401.

Zhao, Z. H., Chang, Y. K. and N'Guérékata, G. M. (2011). Pseudo-almost automorphic mild solutions to semilinear integral equations in a Banach space, *Nonlin. Anal.* **74**, pp. 2887–2894.

Zhao, Z. H., Chang, Y. K. and Nieto, J. J. (2013b). Square-mean asymptotically almost automorphic process and its application to stochastic integro-differential equations, *Dynam. Syst. Appl.* **22**, pp. 269–284.

Zhou, Y. (2016). *Fractional Evolution Equations and Inclusions: Analysis and Control* (Elsevier, New York).

Zhou, Y. and He, J. W. (2021). Well-posedness and regularity for fractional damped wave equations, *Mona. Math.* **194**, pp. 425–458.

Zhou, Y., Wang, J. and Zhang, L. (2016). *Basic Theory of Fractional Differential Equations* (World Scientific, Singapore).

Index

A

ω-antiperiodicity, 13
asymptotically Bloch-type periodic, 16

B

Banach fixed point theorem, 11
Bloch-type periodic, 13–14
Bochner integrable functions, 2
Bochner theorem, 2

C

C_0-semigroup, 4

D

dominated convergence theorem, 3

E

E-resolvent operator, 100
eventually norm continuous, 5
exponentially bounded, 5

F

fractional differential equation of
    Sobolev type, 99
Fubini theorem, 3

I

immediately norm continuous, 5, 65
integral resolvent family, 154

K

k-regularized resolvent, 9
Krasnoselskii fixed point theorem, 11

L

Laplace transform, 7
Leray–Schauder alternative theorem,
    11

M

mean value theorem, 4
measurable function, 1
Mittag–Leffler function, 70

P

ω-periodic, 31
Pettis theorem, 2
point spectrum of A, 4
pseudo S-asymptotic ω-periodicity,
    14, 31

pseudo S-asymptotically
    ω-antiperiodic, 37
pseudo S-asymptotically Bloch-type
    periodic, 37
pseudo Bloch-type periodic, 17–18

R

α-resolvent family, 114
(α, β, γ)-regularized family, 129
resolvent of A, 4

S

p-Stepanov bounded, 48
S-asymptotic ω-periodicity, 14
S-asymptotically ω-antiperiodic,
    31–32
S-asymptotically ω-periodic function,
    14
S-asymptotically Bloch-type periodic,
    32
S^p-weighted pseudo S-asymptotically
    Bloch-type periodic, 54
sectorial operator, 6
sectorial with respect to E, 100
Sobolev-type resolvent family, 100
Stepanov-like (weighted) pseudo
    S-asymptotically ω-anti-periodic,
    49

Stepanov-like (weighted) pseudo
    S-asymptotically ω-periodic, 49
Stepanov-like pseudo
    S-asymptotically Bloch-type
    periodic, 48
Stepanov-like (weighted) pseudo
    S-asymptotically Bloch-type
    periodic, 48
Stepanov-like (weighted) pseudo
    S-asymptotically periodic function,
    14
strongly continuous resolvent
    operator, 146

U

uniformly exponentially stable, 5, 62
uniformly integrable, 5

W

weighted pseudo S-asymptotically
    ω-antiperiodic, 45
weighted pseudo S-asymptotically
    Bloch-type periodic, 45
weighted pseudo Bloch-type periodic
    functions, 20, 22

Printed in the United States
by Baker & Taylor Publisher Services